Humble Heroes

The Middle Fork's Anderson Family

PATRICK TAYLOR

Copyright © 2020 Texas Yeti Books

All rights reserved.

Photos & images available at:
http://www.texasyetibooks.com

Cover design by Kristin Bryant
Kristindesign100@gmail.com

ISBN: 9798657233803
ISBN-13: 978-9798657233803

***

Other Books by Patrick Taylor:

"Lost on Purpose"
"River Hippies & Mountain Men"
"Alone on Purpose"

CONTENTS

Acknowledgements	vii
Humble Heroes	1
Before the War	11
After the War	21
Early Days	35
Terry Remembers	45
Salmon-Challis	51
Changing Times	67
River Manager	83
Big Shots	99
Nowadays	107
Shared Joy	119

Acknowledgements

Many people contributed to the telling of this important tale…

Idaho historians Johnny Carrey and Cort Conley provided historical background in their extensive writings, and motivated me with their obvious passion for the history and people of Idaho.

I want to thank Laurii and Gary Gadwa for offering the archives at the Sawtooth Interpretive and Historical Center in Stanley, Idaho for research for this book. Laurii managed the facility and set a high standard for professionalism and accuracy. Gary served as a Conservation Officer at Idaho Fish and Game, and shared many personal experiences about the Andersons and their contributions to the Middle Fork backcountry community.

I want to thank Tammy Anderson Rice, the lady who championed this project, and who spent long hours organizing and assembling one-of-a-kind documentation. This book was her idea, and she was the driving force behind its publication.

Most of all, I want to thank Ted Anderson, who gave me all the time, insight, and anecdotes needed to make this story real and tangible. He was my partner on this journey, and became my good friend.

"The river delights to lift us free,
if only we dare let go.
Our true work is this voyage,
this adventure!"

- Richard Bach

Humble Heroes

In early hours, the trail down Camas Creek that connected the Bar-X Ranch to the Middle Fork was a fourteen-mile ride through damp cold and darkness, even in summer. Shadowy, seldom exposed to direct light, the slopes surrounding the drainage seemed to block out the sun, making it a chilly and eerie place. Back in the early days of settlement in the area, the Camas Creek trail was the last option for travel from Three Forks to the Middle Fork of the Salmon River. There were few trails in the Idaho Primitive Area as remote and dangerous as the one alongside the violent tributary named for the flowering food source popular with the Sheepeaters. It was too narrow to trail cattle down, and only a few prospectors and backcountry homesteaders used it to travel to the river from tiny mining communities in the backcountry.

The first few miles downstream from Camas Creek's confluence with Silver Creek ran close to the water, where willows and alder grew up in the face of rider and mount. Water from the swollen creek lapped up on the trail which snaked through the pole pines and fir trees that grew up the slope from the creek. The salmon-rich water tumbled and crashed across the rocks, and high water piled trees up like broken barricades in the waterway. It was a preview to the wild and scenic Middle Fork of the Salmon, where Ted's father waited.

As the canyon narrowed and feeder creeks named Hammer, Anvil, and Forge contributed more water to its flow, Camas grew louder, darker, and more violent. Not quite halfway to the river, Dry Gulch opened on the right, a sloping passage to an area known as Yellowjacket, a forbidding craggy cliff zone that served as home to one of the bigger backcountry mines from the Gem State's past. Dry Gulch was the only place on the trail where a packer could safely stop and check his loads. The rest of the trail was too narrow and too steep to allow the packer to maneuver around his string, and too eerie to expect the string to stand still. Two miles further, the trail moved up the steep slope past Yellowjacket Creek, and the rocky distance between the sketchy pathway and the water several hundred feet below made the danger of the ride too obvious; a wreck on that section would be nothing but bad news. And no matter how good a packer was, wrecks were inevitable on the hostile trail; a load would roll or hornets attacked, or some other thing to spook stock. Pack animals would be broken or killed in the fall down the rocks to Camas Creek, and likely the packer would, too. It was hard work, stressful; old-fashioned labor that split finger nails, cracked hands, and frayed the nerves of full-grown men.. It was even more difficult for a teenage boy. He had to have a good reason to make the trip, especially to make it more than once.

"I lost my hat there once," he recalled. "I was daydreaming, and a gust of wind came up behind me and blew my hat off onto the trail up front. I was on Chief, a good enough horse when things went well, but he didn't like seeing that black hat in the trail. He spun around up there just by that Yellowjacket, turning right back into the string. You know where I'm talking about?"

"Yes, sir. I know exactly where. There's no room up there."

"No, there isn't," he affirmed matter-of-factly. "I was lucky to get him turned back around without makin' a mess of things."

Ted Anderson was a young man when his father came up with the idea of repurposing Navy rubber rafts for recreation on the Middle Fork of the Salmon River, dubbed 'the River of No Return' by Lewis & Clark during the crossing of the Rocky Mountains one hundred forty years before. Andy Anderson and his brother Joe ran an outfit headquartered at the Bar X Ranch on Silver Creek, just four miles upstream from Meyers Cove (formerly known as Three Forks). They moved to the Idaho Primitive Area in the late '30s, started out as packers, and eventually realized their ambition of operating a hunting and fishing outfit in the beautiful and rugged backcountry north of Challis. Ted's job was to pack the rafts by horseback down Camas Creek to Andy and Joe, who were waiting with clients a few miles up the Middle Fork at a place called Tappan Ranch. Andy was equal parts mountain man and entrepreneur, and looked for ways to add value to the hunting and fishing expeditions he hosted for sportsmen in search of the ultimate wilderness experience. Andy believed that rafting clients down the Middle Fork would be a profitable supplement to his outfitting business. He did not have the whitewater experience of the Oregonians or the Utah crew that first ran the intimidating river, but he had the motivation and drive to open it up to the general public. World War II was over, and his vision for offering the river to fishermen of means had merit. Andy felt the time was right for kicking off his new adventure, so he acquired surplus military rafts, packed them by horseback to the river, and became the first local outfitter to charge sportsmen for whitewater hunts and fishing trips on the Middle Fork. Andy's enthusiasm was contagious; he was a natural-born salesman, and his business associates and customers bought what he was selling.

Ted, however, was less impressed; a young man in his late teens, he had better things to do than load bulky river rafts on pack animals, guide them down a hostile trail for delivery to Tappan Ranch, and ride a long day back to prepare for the next trip. He went to school in Salmon, a small town of about 1500 on the other side of Williams

Pass. He liked school, especially the social side of it. He played basketball when we could, and enjoyed doing the things teenage boys did in those days. He played when he could, but mostly he worked. He contributed, like the rest of the family, and motored his brakeless 1929 pickup truck to town whenever he got a chance.

"We used surplus survival rafts in those days," Ted told me over lunch at the Y Inn in Challis. "We kept our food in metal honey cans with snap-on lids, and rolled our bedding in canvas to keep it dry."

"When was your first trip down the river," I asked, stumbling to make conversation with a man I considered a backcountry legend.

"June of '45, I believe. Maybe '46. My first time down was when we were asked to search for an old miner who drowned at Survey Creek. We never did find him, but I remember the trip. It was my first trip, but my dad's second time down the river."

Ted was almost ninety years old when I met him the first time. His daughter Tammy was a friend of mine and she invited me to his home. He lived in a small house on the main street of a little town in Idaho, a short walk from the Forest Service facility that served as his base of operations while he worked as River Manager for twenty years. His home on the corner of Tenth and Main was a clean and tidy place, nondescript except for the huge pine tree in the front yard. Tammy had scrap books laid out on the dining table, and was anxious to share memorabilia that accounted for decades of family history. She read books I had written about the Idaho backcountry and wanted me to write a book about her grandpa. There were lots of books about the men who pioneered the iconic river, but nothing more than notes about the family that developed and managed the business of Middle Fork whitewater for generations. It was important to Tammy; not as important to Ted. He loved the river and spent the majority of his long life working on it, but didn't see where his contribution had been any greater than anyone else. He listened as Tammy related anecdotes about her father and grandfather, and he added or corrected details when appropriate. Both her brother and husband contributed to the conversation, and everyone stood around the serving counter that separated the kitchen from the dining room. There were pictures from the Bar-X, advertising pamphlets from the 1940s, and magazine articles from Life, Field & Stream, Argosy, and the Saturday Evening Post. I was surprised at the press clippings; I had no idea that Andy received so much publicity, or that it was national in reach. Ted's kids – in their fifties and sixties - paged through picture books and retold stories each had told a dozen times or more. Their pride in family was evident, and rightly so; theirs was a history that now spanned four generations.

As the conversation continued, Ted moved toward a chair at the dining table, and readied himself like old men do before sinking down to take a seat. But as he turned his back toward the chair and began to bend his legs, he made eye contact with me, then seemed to change his mind. With an imperceptible nod of his head and a 'come this way' wag of his finger, he rose from almost-sitting to lead me through the living room. I paid attention as I followed him toward the hall. He wore a blue-on-blue plaid flannel shirt open for the first two buttons, and jeans. He still had hair, but not much of it, balding in the middle on top. His glasses were large copper-colored metal ones, which framed brown eyes that were very much alive. He felt familiar, like an uncle might, and I followed him to the short passage from the main area of the house to the bedrooms.

"This is my Hall of Fame," he said matter-of-factly, pointing at pictures that lined both sides of the hall. Two dozen pictures in wooden five-and-dime frames adorned the walls around us.

"That's Bush. Bush Senior, with the Department of Agriculture Secretary and myself."

I knew the former President's face, but he was younger in the picture than I remembered. Set in the lower left-hand corner of the frame was a wallet-sized snapshot of Ted with some goofy Groucho Marx glasses, a contrast to the straight-faced tour guide in front of me. And it made me smile a little; clearly, he didn't take himself too seriously.

"And this over here, in this picture, is Peanuts… that's what I called him. Carter."

Another former President of the most powerful nation on Earth.

"..and his helicopter." Signed and personalized for the river guide. Each picture of dignitaries was also a picture of Ted Anderson, stocky and wearing a life vest, with a Forest Service cap on his heavy head. He seemed unimpressed, as if sharing his history was more a duty than a pleasure. If Ted had an ego, it did not accompany us to the hall.

He glanced at other pictures on the wall, but did not bother to explain them. Instead, he turned around and pointed at frames on the other side. He looked back and forth at the two color portraits that were faded brown and amber and auburn by long years in dim light in the hall.

"That's my dad. And me." He looked with his chin up and put his hands in his pockets.

The pictures of father and son were the ones I remembered most. They were full frontal portraits, both men straight-faced, and both wearing red bandanas on their heads, not to be cool, but to catch the sweat that came from long days in the sun. They wore white t-shirts and shared some facial features, though Ted's unshaven face had darker hair and eyes. Andy looked tired, like a man who stayed up too many nights planning and reworking his plan for success. Ted looked thick and hard, his chin square like his father's, but his ears were bigger and his eyes were without bags. There was a hint of a smile on Ted's face, which told me even more about his dry sense of humor. I stared at the pictures to get a feel for the men.

There was a picture of a baby, maybe his daughter, on the wall. He pointed at a group in a raft splashing through some rapids.

"And this is my family, on a family trip. Those were my favorites."

His favorites, he said. Not the ones he took with Presidents and VIPs as a representative of the Forest Service, but the trips he took with family.

He paused, enjoying what seemed a pleasant memory.

"I just thought you might be… yeah," he finished without finishing, a modest man sharing what someone else found interesting. He turned off the light and led me out of the hall, back to where the family was gathered around magazine articles and photos.

I first became familiar with the Anderson family when I hired on as a packer at the B-C Ranch, located on the very same site and in the same buildings as the Bar-X Ranch once owned and operated by Leon 'Andy' Anderson. I worked for an outfitter who leased the facility from a wealthy farmer that bought the ranch in 1997. The outfitter was interested in its history, but did little to preserve the historic site, and it degenerated in his care like the granite slopes that surrounded it. When I came in, it was a 'hill jack junkyard'; the bones of three snow machines littered the space between the old lodge and the creek that ran along the western edge of the property, and unremarkable collections of wood and metal filled the yard behind the building that once served as Andy's lodge. The roof leaked and fence posts leaned on rotten jacks. Still, it was recognizable as the ranch Andy and his brother Joe featured in brochures and magazine articles as headquarters for their trendsetting outfitting business. The Bar-X Vacation Ranch was a piece of Idaho history, and I was instantly captured by the story behind it. I marveled that I learned to pack on the same ranch where the Anderson's started a backcountry industry.

Being from Texas, I had never heard of the Anderson family and knew nothing about whitewater recreation on the Middle Fork until I started working at the B-C. All I knew about Idaho was that it was one of the last primitive areas of the country, where hard men harvested big elk and hosted hunters with money and bravado enough to vacation in its dark steep mountains. I knew that the ranch bordered the Frank Church Wilderness – the largest contiguous wilderness area in the Lower 48.

I did not know that the Middle Fork was the second-most-popular whitewater destination in the country, with ten thousand visitors floating the river every summer. Thirty-something river outfitters provided every kind of backcountry camping experience – from roughing it to 'glamping' - for those interested in the wild and scenic river trip, and dozens of private groups floated in the permitted spaces remaining. The only destination more popular than the Middle Fork was the famous Grand Canyon, although the Impassable Canyon at the end of the Middle Fork boasted higher vertical relief. Reading about its history while working at the ranch, I was amazed at what I did not know.

It seemed so long ago, but in truth, it was the recent past; just a couple generations after Civil War veterans populated Leesburg, Idaho in an effort to 'strike it rich', and less than fifty years after the Sheepeater War. When mail was delivered in the winter by a sled driver behind a team of dogs, and they experimented unsuccessfully with carrier pigeons to communicate from backcountry camps to the lodge. They pioneered their dream before I was born, before the Second World War, before virtually anyone in the country even knew what 'whitewater' was… and only a handful of years after the best whitewater men in the West successfully ran the river in foldboats, and wooden boats with conveyer belting tacked onto the bottoms. Before there were roads to put-ins and pull-outs, when there was no such thing as Search-and-Rescue. Back when gas was a couple dimes a gallon, and eight hundred bucks bought a new car. When Hemingway fished the river and Zane Grey packed to Thunder Mountain. On the tail end of World War II, the Andersons loaded their first paying customers onto rubber rafts and floated them down the Middle Fork for the wilderness experience of a lifetime.

The Bar-X Ranch was located on the north side of Silver Creek, about ten miles southeast of Rabbit's Foot pass, on the edge of the Middle of Nowhere. It was a little more than an hour's ride on horseback from Meyers Cove, a place of great importance to the backcountry community that supported early mining activity in the area. In the early 1940s, the Bar-X became the centerpiece of Andy's hunting and fishing outfit. But there was little to suggest it would produce champions of whitewater recreation in Idaho.

History seemed to favor the bold and adventurous, and young men who thought themselves bold and adventurous carved their names into trees and left their sign by the trail in vain attempts to be more than they were… to create legend where it did not exist. In reality, history was more often made by unassuming men whose secret accomplishments seemed the byproduct of extraordinary efforts to lead ordinary lives, to provide for their families, and maintain lifestyles that were more important to them than fame or financial success. Middle Fork history was made by the Andersons, a typical American family that, over a period of almost sixty years, helped create and manage one of the largest and most unique wilderness experiences in the country. They accomplished things of lasting importance that few in their community would ever know about. They were heroes forever humble for whom we should be thankful.

Before the War

At the beginning of the twentieth century, the Middle Fork of the Salmon River was virtually unknown. Although home to Sheepeater Indians for centuries, the first white settlement in the area was not established until 1867, when Yankee and Confederate veterans of the Civil War flocked to make their fortunes in the gold mines along Napias Creek. It was decades before anyone paid any attention to it at all. It wasn't until 1885 and the development of a mining camp at Thunder Mountain that miners and packers were forced to negotiate the Middle Fork, and then only to cross it, to deliver necessary supplies.

In time, ranches were established on the small flat pastures that dotted the river. The Beagle Brothers raised cattle downstream from Camas Creek, and John Ramey homesteaded the land at the confluence of Loon Creek and the Middle Fork in 1916. He heard about the land from his uncle who fought in the Sheepeater Campaign of 1897. Before that, only lonesome prospectors and Chinese miners bothered with the remote backcountry, more prosperous mines having been found on the eastern side of the river. People died crossing the Middle Fork. It was too far from towns to be of much use to anyone but those supplying mines with cattle and way stations.

It was not until twenty years after the turn of the century that any recreational activity on the river was documented. Running 'whitewater' was an unknown pursuit to the general public, although it had fostered enthusiasm on the East Coast and among a very few in the West. Many of the great rivers had seen attempts at navigation by names unknown to most people; icons in early whitewater boating, and some of them made history. In 1869, Major John Wesley Powell surveyed the Green and Colorado Rivers, and commandeered the first crew to successfully navigate them all the way through the Grand Canyon. Nathaniel Galloway began making boats specifically for running the Colorado River through the Grand Canyon as early as the 1890s, and made his first run in the early 1900s. Beyond the Green and Colorado, 'whitewater' in the West was mostly run in Oregon. It was the domain of adventurous souls who identified with things wild and remote, and it did not catch on with the American public for another generation.

In 1915, though, Cal Allen said to Glen Wooldridge, "Let's make a boat and go down the Rogue River", and that may have been the birth of whitewater recreation in the West. Oregon was coastal, and much of its population grew up on or around the water. It was natural for them to turn to its wild rivers for recreation and, by the 1920s, for recreational business. Celebrities from across the country sought out Wooldridge for guided whitewater fishing trips down the Rogue and other destinations in the state. Some of the big names in Oregon whitewater made their way to Idaho to run the Salmon and Snake Rivers. While they lay claim to taming many of the big rivers, they were not the first to run the Middle Fork.

Henry Weidner lived in Payette, Idaho, and made sure that the first man down the Middle Fork was local. He was an enthusiastic hunter and fisherman, and had run the Main Salmon twice; once in 1911 and again in 1921. He seemed fascinated by the rough water he observed in the Middle Fork at its mouth as he passed by on the

Main. Since, to his knowledge, it had not been run, he found it too good an opportunity to ignore. By 1926, he planned and executed a three months-long expedition down the river, taking two canoes, his son, and two companions. Primarily, it was a photography trip; Weidner was a talented camera man and motion picture enthusiast. His efforts to 'make it big' in the movie business came to nothing (if not worse), but he did make it down the Middle Fork, and claimed the prize for Idaho.

It is possible that another Idahoan named Harry Guleke ran the river in the '20s, but his story is sadly suspect. He was, in fact, a noted sweepboat pilot on the Main Salmon and was sure to have passed the mouth of the Middle Fork many times while guiding his scow down the big river. Historian Cort Conley said in "The Middle Fork Guide" that Guleke may have run a portion of the river, but probably only the last twenty miles.. and few question the learned opinion of Mr. Conley. In absence of any historical documentation or other meaningful evidence, the run is considered more fiction than fact.

Perhaps the group that best represented the early spirit of the Middle Fork was a cadre of white water enthusiasts from Utah that called themselves the 'Colorado River Club'. Led by a sheriff named Frank Swain and a carpenter named Bus Hatch, the club started in Vernal when the two cousins got caught up in the stories of a prisoner named Parley Galloway. Parley was the son of Nathaniel Galloway of Grand Canyon fame. Parley had no money for bail and no prospect for getting any, so he entertained the sheriff and his cousin with whitewater stories from his youth, and promised that he'd teach the men how to build their own boat if they let him out of jail. When he gained his freedom, however, he reneged on the deal, and skipped town before making good on his promise. Frank and Bus didn't let that stop them; along with a group of friends, they developed their own boats and techniques, and (in the words of Bus' son, Don) they "probably set river running back twenty years".

Like most of their efforts, the organization of their club met with difficulties. They started their hobby at the worst possible time; at the beginning of the Great Depression. There was not a lot of work in Vernal and, consequently, not a lot of money to spend on equipment or adventure. But Swain had taken a job as a security officer at Utah Copper Company south of Salt Lake City. He had the good fortune of meeting Russell 'Doc' Frazier there, who was working as the company doctor. Frank got Doc on board, and Doc brought along his friend Bill Fahrni, who owned the supply store near the mine. Both men joined the club, which was good luck for the rest of its members because both men had money and financed many of the group's excursions.

Throughout the early '30s, they ran all the famous (and infamous) sections of the Green and Colorado rivers. They were rough and ready for action; devil-may-care adventurers with boats named "What Next", "Who Knows", and "Don't Care". Through trial-and-error (an abundance of error, as their history makes clear), they persevered, and managed to build a rowdy reputation as river runners. In 1935, they made their play for the Middle Fork.

"It was like riding a wheelbarrow down a flight of stairs," remarked Royce 'Cap' Mowrey, brother-in-law to Bus Hatch. Their first attempt in 1935 was thwarted by the low water and rocky terrain. With help from local packers, they cached their boats at the bottom of Dagger Falls (just before the present-day put-in point at Boundary Creek) to pick up when they returned the next year. They reconsidered their boat design, and implemented improvements for their next attempt (they screwed and tacked conveyor belting to the underside of their boats). They realized success on the second trip, running from source of the river to its mouth in six days in the summer of 1936. (It was during their second trip down the river that they discovered the presence of the hermit Earl Parrott, who hid from them at the river's edge.)

Doc was hooked. He wanted to run the river again in '37, so Bus and Cap built two new boats. But they were busy with long-awaited work, so they couldn't go along. Unfortunately, the new boat design failed, and Doc left the boats in disgust to rot on the river's edge. He tried again in 1939, and was successful. Fahrni went along, as well as Swain. Other newcomers included experienced river pilot Amos Burg (which is noteworthy because he was a pioneer of rubber rafts) and he made the run in a custom rubber raft made by Air Cruiser Company. Burg ran the raft on the Green/Colorado run the previous year with whitewater legend 'Buzz' Holmstrom and a man they picked up along the way, Willis Johnson. Burg and Johnson both joined Doc's group for the '39 Middle Fork run. The trip took 12 days, which included a stay-over with Earl Parrott in the Impassable Canyon. During those four years, Doc became the defacto expert on running the Middle Fork of the Salmon.

That same year (in fact, only three days after the Colorado River Club's run), a Texas transplant to Oregon named Woodie Hindman took his wife Ruth on a private fishing trip down the river he spotted a month earlier while guiding a fishing trip on the Main Salmon. They floated a modified McKenzie boat that he built in Eugene, Oregon. To everyone's surprise, Woodie and Ruth were able to run the river alone. The next year, Woodie brought along three friends, effectively introducing Oregon's whitewater elite to the magnetic Middle Fork of the Salmon.

It was during this time that Andy Anderson entered the picture, after he moved to Idaho and started his backcountry outfitting business.

'Andy' was born Leon Leroy Anderson in Bothwell, Utah in 1910. His grandfather was a Danish farmer who converted to Mormonism and immigrated to the United States in the early 1880s. His father Lawrence Leroy worked for the United States Bureau of Fisheries, and the family lived near the border of Utah and Idaho. Theirs was a subsistence lifestyle, and every day was a working day spent in the high desert mountains and forests of northern Utah. Andy, a nickname given to him by his friends, was the oldest of five boys and the smallest of the group. He worked with his father and learned at an early age all the outdoor skills that served him the rest of his life. As a young man, Andy ran a trapline for bobcat and coyotes. He earned $12.00 for each bobcat caught and $5.00 for each coyote.

"Not bad money for a high school kid… in those days," he later remarked.

He spent a couple years riding and roping in rodeos in Wyoming, Montana, and Idaho. Accidents and injuries left him with a tricky back and a broken leg. He wore a built-up shoe the rest of his life to accommodate the shorter limb. Afterwards, he played left field and pitched for the Utah Copper Company, a semi-professional baseball team. A scout for the Oakland Oaks (a Pacific Coast League team) liked Andy's style and offered him a contract to play professional ball.

"Maybe I should have signed, but the wilderness had a stronger pull on me than the city lights," he was quoted as saying years later. It was a statement that said much about the patriarch of the Middle Fork's first river family. Not many men would forego the opportunity to play a game for a living, but Andy knew his heart, and stayed with the life he loved.

Later, he worked for the Federal Fish & Wildlife Department, and was recognized as an expert in predatory-animal-and-pest control. During those eight years, he managed the Minidoka Migratory Waterfowl Refuge on the Snake River. Andy's contributions to the backcountry began early in his life.

He married Melba Fink in his late teens, and fathered Ted Anderson in 1929, at the beginning of the Great Depression. Two other children were born to Andy and Melba; Connie and Terry completed the young family. Although the country's economy faltered and many families suffered, there was little change in the Anderson household. They were outdoorsmen, mostly living off the land, and the land changed little during the Depression. Andy found work that suited him, work that fed his family and kept him out-of-doors.

In the years leading up to World War II, Andy raised beef cattle and worked as a welder for the Remington Arms Company in Salt Lake City, Utah. He was deemed unsuitable for service (classified '4F' by military doctors) because of an eye injury suffered as a child, the result of staring at the sun through a hole in his straw hat. During that time, the Anderson family built the foundation for the success they would know in post-war years.

In 1938, Andy moved his family to the Lemhi area and established a packing outfit at the mouth of Warm Springs near Meyers Cove, deep in the Idaho backcountry. He started out with a dozen head of horses. The first job for his pack string was packing supplies for a forest fire on Little Bear Creek. He and his brothers Joe and Cy packed bread, eggs, and other goods and headed for the fire. He made frequent trips to the primitive area in Central Idaho (that decades later would become the Frank Church Wilderness) to guide occasional customers on hunting or fishing expeditions. He and Joe set up a camp off of Silver Creek in an area known then as Warm Springs (later called Spring Creek, just up the dirt road from Meyers Cove). Little by little, the outfit provided enough support for his family's needs, and kept him active in the way-of-life he enjoyed most.

Andy loved the Middle Fork country, and learned more about that untamed wilderness than any man alive at the time. He ranched in the heart of it, fought fires for the Forest Service, guided and prospected for gold in it, and backpacked into unmapped sections just to find out what lay beyond the next rocky ridge. Whenever possible, he explored the Idaho Primitive Area, and became familiar with places well off the beaten path. His curiosity led him to Cache Basin, a watered bowl high in the mountains west of Meyers Cove. It was the hub in a mountain range whose spokes led to such landmarks as Falconberry, Tappan, Martin, Woodtick, and the West Fork of Camas Creek. He and his brother had a place where clients could stay during high country fishing expeditions. He cleared trail down the ridges to the river, and scouted elk and mule deer in the ranges that surrounded the high country hub. Andy spend weeks in the Bighorn Crags, an area so remote that locals seldom spent the time necessary to explore the dozens of pristine alpine lakes that were strung among the peaks and pinnacles like pearls on a necklace. It was a place that would become a trademark of Andy's backcountry expertise; other men knew the Middle Fork, but no one knew the surrounding wilderness better than Andy Anderson.

It was about this time that Idaho Magazine opened its doors as the first big press publication in the state. They wanted to legitimize their monthly magazine, so they hired a staff of professional print people and outdoorsmen to attract sportsmen from around the country and build an industry for them in Idaho. By that time, Andy was well-connected in the Sun Valley area, and the editor selected him as the magazine's 'official guide'. It was the beginning of a long, productive relationship with the press that Andy leveraged to build his reputation and his business.

Andy developed sales skills that separated him from other sportsmen and guides. He learned to articulate his vision in a way that appealed to those who longed to experience the outdoors in a manner reflective of their dreams. He learned to describe the wilderness and the experience of living in it, and the pleasure derived from a temporary engagement with primitive living that was available to everyone. Andy was able to craft a dialog that made an individual feel as though he was part of an exclusive club, a fraternity of those blessed with an understanding unavailable to those outside the experience. And the way he explained it made the listener feel that no one but Andy Anderson could deliver on that promise. He became more than an 'official guide'; he became an active evangelist for the wilderness of Idaho.

When World War II began, the Middle Fork was beginning to attract attention as a whitewater destination. The Oregon guides returned in 1941 to become the first to take paying customers on the river. And in 1942, a young college student from Yale named Eliot DuBois became the first man to solo the Middle Fork. He completed the run in a foldboat, an interesting design that was portable yet sturdy, pioneered in Europe and popular on the East Coast. His achievement was noteworthy because he started out as a party of three, but wrecks during the first four miles of the trip disabled his companions' crafts. Eliot was not to be deterred, and made one of the most impressive (if not crazy) runs logged during that time. Eliot then joined the war as a Marine, and most every other river runner had to wait until the war was over to resume their adventures on the Middle Fork.

After the War

Andy and Joe continued in the packing business until Joe went off to war in '42. In the meantime, Andy purchased the Max Oyler place at the confluence of Silver and Ramshorn creeks near Meyers Cove. The ranch would become the headquarters for his outfit, and he named it the Bar-X Vacation Ranch. For the next few years, he worked the Idaho Primitive Area and took guests for hunting and fishing trips miles in every direction. He hunted the Bighorn Crags, fished in Cache Basin, and leased ranch land to use during his hunting and fishing trips. He made friends with Bob Simplot, whose brother Jack was known throughout the state as a big-time farmer and money man. Bob bought the ranch that sat along Loon Creek and the ranch run by Daisy Tappan at Grouse. He traded Andy the rights of use for work on improvements to the properties. Ted Anderson spent lots of time irrigating the Simplot Ranch (as the Loon Creek property came to be called), and he and his father met Hatch and his friends when they made runs down the Middle Fork.

In 1945, Andy finally met Frank Swain, a chance encounter that was to change Anderson's life forever. In the summer of that year, Andy was working the ranch at lower Loon Creek when Frank stopped to camp during another run on the river. Swain's group stopped on the sand bar at the confluence of Loon and the Middle Fork, at the base of the 120-acre ranch. It was a small group of people, which included Swain's fourteen-year old son John, who was crippled from polio.

"I'd like to get a horse for John to ride the next few miles," said Frank. "The boat hurts his back. Would you care to ride with us and pick up the animal when we make it past Tappan?"

"Holy jumped-up Moses, yes!" Andy immediately replied.

The next day, he floated the Middle Fork from Loon through Tappan with Frank Swain and his crew. He was fascinated by the stocky, gray-haired lawman as he navigated through swirling rapids and avoided treacherous rocks. By the time they reached Camas Creek, however, the granite teeth of the river chewed the bottoms of two boats badly, even though they were reinforced with conveyor belting. Andy noted that, while the wooden crafts took a beating, the little yellow life raft, filled with supplies and tied behind the last boat, bounced off the rocks and spun lazily through swift currents, and reached Camas Creek unscathed.

"Why couldn't a man run the river in one of those rubber life rafts?" Andy wondered out loud.

"He could," said the veteran whitewater expert. "What's more, he could develop a business around a new kind of thrill that would attract fishermen and hunters from all over the country."

After he left Frank Swain's party, Andy high-tailed it over the mountains to Pocatello where he found a little yellow survival raft in a war surplus store by the highway. He bought it for fifty dollars and packed it on horseback to the Simplot Ranch on Loon Creek. Shorty Waits, the grizzled wrangler and caretaker at the ranch, thought his boss had lost his mind.

"What do you plan on doin' with that?" he asked.

"I'm going to ride it down the river."

"What do you want me to do with your horses?" Shorty asked with a somber face, uncertain that the man would return.

"Do you what always do. I'll be back in a few days."

Shorty offered Andy a final handshake, as if he would never see the man again.

Like Lindbergh setting out alone across the Atlantic, Andy floated down the Middle Fork by himself in August 1945. Eliot DuBois soloed the river in 1942, but Eliot had experience running whitewater from his adventures on East Coast rivers. Andy had ridden with Frank Swain for about ten miles, and never once took the oars on that trip. It's fair to say that Andy was ill-prepared to make such a dangerous run, and Shorty was not wrong to assume that the ex-rodeo rider was in over his head. But Andy had taken risks before; I guess he felt his bronc-bustin' days qualified him for whitewater rafting. Whatever his logic, Andy pushed off from the beach at Loon Creek and began his first run down the Middle Fork.

When the white water whispered, he let the yellow boat run. But when the river roared and rocks big as houses loomed ahead, he paddled desperately toward the shore. Sometimes, he portaged around dangerous rapids, and sometimes he lowered the boat with a rope. Occasionally, he just rode them out, singing 'Ragtime Cowboy Joe' at the top of this lungs, his voice reverberating off the canyon walls above the thunder of the churning water.

"I was a-swallowing my heart all the way down," remembered Anderson. "Scared of the river? Sure! But I was more scared of what would happen if I broke a leg and couldn't walk out. It would'a been slow death by starvation."

That was a legitimate concern. The problem with the backcountry surrounding the Middle Fork was its remoteness and inaccessibility. It was, by definition, primitive and wild. There were few trails to connect the ranches that sparsely populated the banks of the river, and those few trails were infrequently traveled. Beyond Big Creek, there were no trails at all; for more than twenty miles, until its confluence with the Main Salmon, the Middle Fork was a watery highway without an exit. The rocky walls narrowed into a wild gorge known as Impassable Canyon, whose granite barriers extended almost 4,000 feet above the clear green water. There was no way out except on the river.

"When I got down into Impassable Canyon that first time, I had the lonesomest feeling of my life. Like I was the last man left on earth."

There were lots of rapids in the Impassable Canyon. Eliot DuBois remembered the Beast of the River in that scary section, a roller-coaster of big waves later named Weber Rapids. The Colorado River Club told stories of members getting thrown from their boats while navigating whitewater in the canyon. There were nine big rapids in a 7-mile stretch hidden in the Impassable Canyon. Perhaps the most difficult section of the river - the infamous Hancock Rapids - lay ahead of Andy. They were named after an old sourdough from Salmon City who struggled on foot up the Middle Fork from its mouth and reported the three-mile gauntlet of whitewater as "absolutely impossible to get through alive". It began with a series of foaming rollers called the Washboard. Then, almost without warning, came the Main Hancock, a tricky collection of swirls where huge boulders and a talus slide pinched the river down to a width of ten feet. Then came the Roller Coaster, which tilted a boat almost on its end. If the pilot was not careful, he'd be slammed against a sheer rock wall and the undertow would suck him beneath a wash of bubbles. Next came the Slalom, a tortuous course between shed-sized rocks, and finally the Jump-Off, another half-whirlpool swirl with submerged rocks. Literally, Anderson had nightmares about that part of his first trip for years.

"My first time through Hancock Rapids, I don't remember being scared. I was too busy trying to keep that little yellow boat from folding up on me like a hot dog bun. But I've been paying ever since in my sleep."

After surviving his first trip down the Middle Fork, Andy was sure that rubber boats were the solution to riding the river in relative safety. However, he needed something safer for customers than the little raft he used during his first voyage. The next winter, in a war surplus store in Salt Lake City, he spotted two big black Navy assault boats; twelve feet long, six feet wide, with rubber hides as tough as elephant skin. They had inflatable double bottoms and were compartmentalized, which meant that even with a puncture in one compartment, the boats would stay afloat. He bought them, packed them to the ranch, and tested them out as soon as possible.

His second trip down the river was at the request of a miner's family, and he took Ted with him on that trip. They went to look for the body of a miner who drowned attempting to ford the river on horseback at Survey Creek.

His third trip was in the company of his wife. On September 1, 1945, Melba ran the river with her husband in the little survival raft. She enjoyed the trip, and was happy to have gone along with her husband. That's when Andy knew he could make a business out of whitewater rafting on the river.

Joe Anderson returned from the service in 1946 to rejoin his brother in the packing and outfitting business. Their improved outfit offered high mountain lake fishing, fall hunting, and river trips on the

Middle Fork. They transported their rubber rafts, equipment, and guests by horseback from the Bar-X Ranch down Camas Creek to the Tappan Ranch – a twenty-two mile trip.

They weren't the only ones providing a river experience on the Middle Fork, however. That same year, Hindman and his friend Prince Helfrich brought paying customers in from Oregon. Prince was river royalty, having already established his reputation by being the first guide to boat portions of the Rogue, Blitzen, and Deschutes rivers in his home state. He was a charter member of the McKenzie River Guides Association. He and Woodie were cast in the same mold, and often collaborated in the organization of fishing expeditions. As early as 1941, they guided paying customers down the Middle Fork on fishing trips. Prince set standards, and his family has continued to provide top-notch services to whitewater enthusiasts and anglers for four generations. In '46, Hindman's wife Ruth ran a rubber raft down the Middle Fork and became the first woman to run the river alone. The next year, Helfrich and Hindman began serious work on the river, installing equipment to make launching easier, taking Prince's son Dave on his first trip, and kicking off a business that exists to this day.

Also in 1946, Bus Hatch returned to the Middle Fork. He started a river guiding service he called Hatch River Expeditions to fish and raft the wild Middle Fork. Perhaps no one had more experience on the river at that time, as Bus was one of the original members of the Colorado River Club.

In spite of their river savvy, however, neither the Hindmans, Helfriches, nor Hatches could match Andy Anderson's knowledge of the Idaho backcountry that surrounded the Middle Fork river. He knew intimately the mountains that bracketed the river, the animals that lived along it banks and on the ridges that stood between it and the skyline. He knew the people that had settled it, that worked it for themselves and the mines that supported it. The other river guides could provide a whitewater experience to compete with his offering, but could not provide the wilderness experience that marked the difference between the Middle Fork and other destinations. It was their knowledge and the application of that knowledge toward the management and caretaking of the river and its surroundings that made the Anderson family's contribution unique to the river and the wilderness.

In August 1946, Andy took his first paying guests down the river. He guided Jack Simplot, Leon Jones, and Box Troxal from the Simplot Ranch on Loon Creek all the way to the Salmon River. Andy was more than a great guide; he knew how to leverage his success to bring attention to the opportunities on the river. He had a reporter in Salmon City to interview the famous Simplot on the way back to town.

"Members of the Anderson party said they believed this was going to develop into one of the most popular river trips in the northwest. Its recreational advantages are extremely appealing and with proper equipment it can be made in safety and without undue hardship," Jack Simplot was quoted as saying. It was a statement that made other people sit up and take notice.

Andy's contacts in Sun Valley and at the Idaho Magazine were then leveraged to everyone's benefit. He used Sun Valley as a meeting place for his outfitting trips, and picked up his customers there before transporting them to the Bar-X Vacation Ranch on Silver Creek. He offered 7-day trips, three days of which were spent on the river, and packed customer duffel and supplies by pack string from the ranch to Tappan on Grouse Creek (an area he became familiar with during his outfitting days in Cache Basin). By running six to eight trips each season, Andy made a business out of guiding guests down the Middle Fork. Perhaps more importantly, he promoted the river as a recreational destination nationwide.

Over the next five years, no fewer than twenty magazine articles were published in national magazines naming Andy Anderson of the Bar-X Vacation Ranch as their expert source. A few were local magazines, like Scenic Idaho and Idaho Sportsman & Horseman (1947 and '49 respectively). More important to the industry and the community-at-large were articles that appeared in publications like Life (April 1949), Sports Afield (November 1949), Argosy (August and September 1950), Field & Stream (December 1951), Better Homes and Gardens (May 1952), and Saturday Evening Post (July 1955). These magazines boasted readership in the millions of people, and the advertising value of that kind of publicity was beyond the reach of the local population. Not only did it benefit other river guides and hunting/fishing outfitters, but also hotels, inns, and cabin renters, restaurants, equipment sales and rental companies, local transportation companies, government employees who worked for Idaho Fish and Game as well as the Forest Service, and the many hundreds of Idahoans who worked at these places. Andy and Joe Anderson were the men featured in these stories, but they were not paid directly for their services. Neither were other members of the Anderson family who worked at the Bar-X; Ted as a packer, Terry when he wrangled, aunts and uncles and cousins that cooked, cleaned, and otherwise cared for the people who wrote the articles. In fact, those trips were given to the press at no charge. Andy did not look to reconcile his business balance sheet with income from these articles or contributions from those who benefited from their publication; he wanted only to promote the Idaho wilderness and the recreational opportunities it provided for everyone.

In the summer of 1947, the first promotional article appeared. It was published by Scenic Idaho magazine and written by its editor, Chet Belcher. The title of the article was 'Bighorn Crags'; it was sixteen pages long and featured 32 photographs of the Bighorn Crags, the Bar-X Vacation Ranch, the guides, and the guests. Andy was the official guide for the magazine and referred to in the article as "Daniel Boone of the Bighorn Crags". The feature described the difficulties inherent in vacationing in a location as remote as the Idaho backcountry, complete with overheating vehicles and crank-phone communications. It described accessing the Crags by crossing a slide at the head of Wilson Creek that led to a pass only open two months out of the year. There were no roads to the Bighorn Crags in 1947, and only a few people knew how to get there. The Anderson brothers packed rubber boats for their guests to use while fishing Ship Island Lake, one of 28 alpine lakes in the Crags above 8400'. As if setting the standard for future efforts at preserving the wilderness and its wildlife, the Andersons insisted that their guests fish with barbless hooks. It was a grueling expedition, as described by the author; not an excursion that would be considered a vacation. But it painted a fine picture of the rewards realized when one truly 'experienced' the wilderness. It spoke volumes about the Idaho backcountry and the expertise of its official guide. No one else had the necessary knowledge of the area to put that trip together.

During April 1949, Peter and Helen Brooks of Boise, Idaho brought an army of publicity specialists to the Bar-X Ranch to gather information and experience for a series of articles and promotional films they intended to produce with associates. Included on the trip were Al Brick from Fox Movietone and Loomis Dean, a photographer for Life Magazine (with a total of fourteen cameras on the trip). The first article to develop from the trip came out in the June issue of Idaho Sportsman and Horseman titled 'White Water', the first reference to the Anderson whitewater recreation offering. While circulation was limited, it paved the way for national exposure. The article made a point of introducing the Anderson family to the public; Andy and his wife Melba, good-looking-enough-for-Hollywood Joe, Ted (and his wife Phyllis, who he married the year before while on leave from the Army), and Ted's younger brother Terry. It was a theme that continued for most of the years that Andy was in business; it was a family affair. Twenty-six year old Hank Hastings worked as a camp cook for Andy, but the rest of the ranch was run by members of the Anderson family.

The article described riding down Camas Creek, camping at Dry Gulch, and arriving at Tappan Cabin on Grouse Creek to begin the river trip. The author familiarized the readers with the seven-man rubber rafts used by the family, including modifications made by Andy to make the experience a little more 'guest-friendly'. Mention was made of incidents that highlighted the safety of the boats; they didn't sink even when filled with water (during a portage over Grouse Creek Falls) or when punctured (a four-inch hole in the bottom from a rock near Camas Creek). It was interesting that – even in 1949 – they described having to clean up a mess left at a campsite by deer hunters. The article described eagles in the sky, huge herds of bighorn sheep, unbelievable fishing, songs around the campfire, and lots of delicious food. In a classic example of Andy's endearing entrepreneurship, they stopped toward the end to climb up to a cave and sign the register for the 'Middle Fork Whitewater Club', with only 27 members at that time. They took pictures, had a small ceremony, and delivered membership cards to each participant. Predictably, the authors confirmed "it was a marvelous experience and a wonderful vacation. We recommend it to everyone as most thrilling and enjoyable."

Published at the same time (though dated May 1949) was an article titled 'Shooting the Salmon; It Is the Roughest and Wildest Trip in the U.S.' in Life Magazine. Life was the first all-photographic American news magazine, and, at that time, boasted national sales of up to 13.5 million copies each week. The article used eleven photographs including a full-page picture of Andy negotiating Hancock Rapids. Life was a magazine that catered to a mature

American audience and was popular in every household. It featured educational articles (the cover of that May issue touted a primer on the atom), news-of-the-day articles, and pieces promoting the American way of life. And that was the value of the article on Idaho recreation; it implied that, however out-of-reach the dream of adventure may seem, the wilderness experience was no longer outside the mainstream. Unlike a feature in a sporting magazine, Life's article spoke to the American public-at-large, and the article by the Brooks' opened readers' eyes to an alternative vacation spot that appealed to the post-war public.

Five months later, Sports Afield published an article titled 'White Water' detailing the Bar-X whitewater vacation. It was written by Bob Reilly and George Long, Jr. and included sixteen photographs. Like every article before it, 'White Water' featured a portfolio of action shots to grab the readers' attention, and it worked. Words accurately described the events, but the thrill of adventure was better communicated with professional photographs taken by enthusiastic participants. In fact, there were more columns of photographs than of print in the piece. Sports Afield was the oldest continuously published outdoor magazine in North America, on the newsstands since 1887. It had a circulation of 800,000 monthly and focused on the outdoorsman. Sports Afield was the first national magazine catering to sportsmen in particular that featured Andy Anderson's tempting offer of whitewater recreation in the wilderness.

Peter and Helen Brooks were from Boise, Idaho, and their interest in Andy's adventure offering resulted in more than a professional connection. They returned many times to the Idaho backcountry, converts to the new whitewater holiday. And they weren't the only writers to take an interest in the Anderson passion. Cecil 'Ted' Trueblood made several trips with Andy and Joe, and through the years became a family friend. Unlike other authors who wrote about the outdoors, Ted was an outdoorsman who liked to write. In 1937, he became a reporter for the Deseret News in Salt Lake City, Utah. From there, he began writing articles for Field & Stream, where he later became fishing editor of the magazine in New York City. In 1947, he moved back to Idaho in order to "fish, hunt, and write about it." From his home in Nampa, Idaho, he remained an associate editor and contributor to Field & Stream and continued writing articles for the magazine throughout his life. It was about this time that Trueblood and Andy Anderson became friends, as both were enthusiastic outdoorsmen and conservationists. In addition to his magazine work, he also wrote several books about the outdoors.

A June 1950 article in Outdoorsman told how Andy donated his time and a pack string to assist Idaho Fish and Game with stocking 20,000 brook trout fingerlings in four lakes in the Bighorn Crags. It was another example of that which set Andy apart from other outfitters in the Idaho Primitive Area (as it was known at the time); Andy 'paid his dues' with the community. He gave of his time, experience, and expertise to make sure the wilderness would be around for future generations of visitors. He did all he could to guarantee the wilderness experience for others. His ability to provide pack strings and packers created greater opportunity for vacation diversity and differentiated his outfit from others on the river.

'Trophies of the Middle Fork' was published in the December 1951 issue of Field & Stream. With Ted Trueblood's help, Andy's business enjoyed promotion in the famous national publication catering specifically to hunters and fishermen. It advertised itself as America's #1 Sportsman's Magazine. It was an article about the Idaho Primitive Area as "a pack-tripper's paradise for deer, elk, and bear", and detailed a week-long hunt along the Middle Fork with Andy as its hunting guide and packer for its [primarily East Coast] readers. Photographs of long pack strings crossing the river and carrying every kind of trophy animal spoke volumes about Andy's expertise.

Perhaps no magazine provided greater exposure to the general public for outfitters and Idaho recreation than the Saturday Evening Post did in July 1955 when they featured an article titled 'He Tamed Our Wildest Rapids' written by Andrew Hamilton (who wrote 'Ordeal on the Continental Divide' a year later, wherein he told the story of Andy saving three lives one night by walking through a blizzard to get help). While only three pages long with five photographs, the article caught the attention of millions of Americans. By the time the article appeared in the Saturday Evening Post, Andy had navigated the Middle Fork forty-two times and was declared "the one man who has completely mastered its deadly rapids". It was a dramatic piece full of river history, but its focus was Andy Anderson. Fittingly, the most widely circulated article generated by his efforts to publicize Middle Fork recreation used most of its column-inches to tell the story of Andy Anderson.

Although difficult to quantify, the impact of the publicity fostered by Andy Anderson's effort to introduce the vacationing public to Middle Fork recreation was reflected in the number of people reported running the river in the years that followed that early media attention. In 1949, 25 people paid to float and fish the river. By 1965, the number grew to 1200, and increased four-fold over the next ten years. Anyone with experience in marketing knows that it takes multiple impressions to make a sale, and the twenty articles written about Andy and the Bar-X Vacation Ranch made plenty of impressions on people across the country. His contribution to the development of Middle Fork recreation was unique, and part of his personal legacy.

The Early Days

In the beginning, no one needed a permit to outfit on the river. There were only a few guides capable of providing the service, and the Forest Service had no interest in their business.

Andy and his brother Joe, together with Andy's son Ted and other members of the family, hit their stride in the late '40s and early '50s. The outfit grew in popularity and professionalism, and it was reflected in the business value of the company. The first paid trips offered by the Oregonians in 1946 are said to have generated between $160-200 each; the advertised price of Andy's week-long river trips (that included only three days on the river) was $500. It was a lot of money at the time, but Andy and Joe were able to book from six to eight trips each summer, which was not much less than professional outfitters are able to fill nowadays. It was their practice to pick customers up in Sun Valley and drive them to the Bar-X Vacation Ranch deep in the Primitive Area. Once they arrived at their backcountry headquarters, they spent two days readying their guests for the whitewater trip. They had a marvelous facility and could offer more than just a ride in a boat.

The Bar-X sat in one of the only sunny spots in the Silver Creek Canyon. Located just a few miles downstream from the historic

Singheiser Mine, the ranch had the best of the backcountry in its front yard. Ramshorn Creek ran right through the yard, and joined Silver Creek across the road. Guests could fish for trout on the ranch itself, or go for trail rides in the mountains surrounding the log cabin that served as the main lodge. There were mule deer in the hills, and elk could be heard bugling at night. Visitors were enthralled by the proximity of real wilderness, experiencing it for themselves before they even got out of the cars that transported them from Sun Valley to the ranch. Singheiser was a productive gold mine and 'color' could be found in Silver Creek, if a guest were inclined to pan for it. Terry (Andy's youngest son) told heart-warming stories about the days of his youth, which left a lasting impression on him. Andy invited lots of extended family members to join in the fun; they assisted customers while they enjoyed the backcountry. Good meals, campfires, and sing-alongs filled the hours of the first two days before guests watched as their duffel was strapped on pack animals to begin the long trail ride to Tappan Ranch. It was just enough time to prepare them for the next stage of their adventure, which raised the bar of experience for everyone involved.

From the Bar-X Ranch, guests would ride a handful of miles down along Silver Creek to Meyers Cove, where the canyon opened up to a wide, green pasture that was an important cattle ranch during mining days. Silver Creek joined Camas Creek, as well as the West Fork of Camas at that place, which is why it was known as 'Three Forks' at the turn of the century. While tucked up in the canyons, it was hard for visitors to get a good view of the mountains that populated the Idaho Primitive Area. It was characteristic of the area; the slopes were too tall and steep to allow a long-range view, as they literally blocked the line of sight. At best, a rider at trail level could see the shoulder of the mountain in front of him, but rarely caught a glimpse of the peaks that crowded the wilderness skyline. At Meyers Cove, however, the slopes laid back where the three creeks came together, and the real majesty of the surrounding countryside revealed itself. It was literally a breath of fresh air, and one to be cherished… because as soon as they arrived at Meyers Cove, the riders took a right to trail down Camas Creek a total of fourteen dark miles to its confluence with the Middle Fork.

In the truest sense of the term, Camas Creek was a wilderness experience. It had an emotional impact on its visitors, and no one made the six-hour trip without feeling the power of the water rushing alongside the steep mountain trail. It let guests know in no uncertain terms just exactly how remote their trip had become. Never mind that the ranch was a two-hour drive down a Forest Service road from the two-lane blacktop highway ten miles north of Challis; the day-long trip down Camas Creek was like going back in time. There were no 'little problems' that deep in the backcountry, and few escaped the subtle dread that accompanied them on the dark trail ride. Because the canyon walls were so steep along Camas, the sun rarely delivered a full measure of heat to the trail. Even on summer days, there were sections where one felt inclined to don a jacket or a sweater. Bar-X stock was familiar with the trail, and Ted brought the pack string with duffel and supplies separately, so there was little to worry about with regard to their mounts. Nevertheless, the trail left a lasting impression on Andy's guests, and provided a stark transition from the comfort of the lodge to the realities of the wild river ride up ahead.

At the end of Camas Creek, they crossed a pack bridge and rode over wide benches that physically relaxed the group. At the Middle Fork, the canyons opened up again. The afternoon sun greeted the riders as they made the left turn across the bridge and onto the trail that led to Tappan. It was a little more than an hour's ride to the small ranch on Grouse Creek that served as Andy's put-in, and a good part of that ride was on level ground at water's edge. After a couple miles, the trail moved up the left side of the river onto steep rocky ground, but the sunlight made the path less intimidating than the ride through

Camas Creek's canyon. Toward the end, the trail moved up the shoulder of the ridge that sloped down the mountains on the left, then emptied into the pastures that marked the outskirts of Daisy's place.

Grouse Creek flowed down a drainage north of Cache Basin and into the Middle Fork just past the halfway point of the river. Over the years, it became the most popular campground for river runners. M.A. Curtis applied for the tract in 1910, but Willis "Bill" Jones first settled at the spot in 1917. It was roughly 74 acres of good pasture land, a nice spot for growing hay, grains, and fruit trees, which Jones did to proof the patent (which became official in 1927). He sold the place to Fred and Daisy Paulsen-Tappan, and Daisy made it famous. Fred was often gone to earn money they needed to buy the things they could not grow, and Daisy stayed behind to tend to ranch work and raise her two sons. She proved herself as competent as any man in the backcountry, and endeared herself to the locals. She became part of Middle Fork lore in the short time she lived at the ranch, later retiring to nearby Challis. When she left the ranch, she sold it to Bob Simplot. Being a good friend to Andy Anderson, Mr. Simplot let Andy use it as a stopover for his river trips. The overnight stay at Tappan added value to the wilderness experience Anderson offered his customers; it provided a unique transition between the trail and the river, one not available through any other outfitter. There was a wide beach (one of the most picturesque campsites on the Middle Fork) upriver from Daisy's cabin, dotted with tall Ponderosa Pines and alpine fir, and the river turned toward the cabin and flowed past it on its way to the main Salmon. Tappan was one of the prettiest spots on the river, and always a favorite camp site for outfitters and guests alike.

There were few places in the Lower 48 with as dark a sky as Central Idaho. Gazing north, the greater and lesser constellations were visible from the river's edge. All seven stars of the Plough (or Big

Dipper, as it's known in the United States) stood out as bright dots on a black background, and Polaris (the North Star) was easily seen, even early in the evening. The Little Dipper was apparent; though rarely seen from the big cities around the country, its dim stars fell away from the North Star. It leaned toward its big cousin and captured the viewer's attention. East and equidistant from the Plough through Polaris, wilderness watchers measured the wave of lights cast by the stars of Cassiopeia. The stars turned and Time passed unnoticeably as the two constellations danced around the North Star, and modern travelers shared a bit of eternity with ancestors of ancient times. The eons of time that it took the river to carve canyons into the mountains were a fraction of the time it took the light from the heavens to reach us, and everyone who looked out on the night from that particularly dark place in America was able to see something that they probably never saw before. No matter the events of the day, a night along the Middle Fork in the Idaho Primitive Area put people in a pleasantly hypnotized state of mind.

The next day, the boats were rigged and packed; everything was made ready for the ride down the river. It would be the first time guests actually saw the rubber rafts. Gear was stowed in watertight compartments, life jackets were distributed, and orientation completed. The objective was to launch before lunchtime. Sometimes they had to line the boats down Grouse Creek Falls (or Tappan II, as it is currently known), less than two miles from the cabin. Guests could fish during such delays, and often hooked big fish while the guides wrestled with the boats. They rode the water the rest of the day, usually stopping in the area above Bernard Creek near the Flying B Ranch for their first camp on the river.

The second day on the Middle Fork was as leisurely as the first. The guides allowed plenty of time for meals and recreation. They started the day with Haystack Rapids and proceeded downriver, where the topography changed, and so did the water. Instead of the alpine setting that was characteristic of the upper Middle Fork, the middle part of the river was considered a 'high desert', and the land around the waterway opened up. The mountains sat back from the water and the line-of-sight increased. Folds in the mountain range became evident and the high peaks dotted with trees displayed a unique aspect of Idaho to its guests. The ridgelines in the wilderness were jagged and tall, but not tall enough for a tree line. There were trees decorating even the tallest ridges, although scattered and sparse in places. Sage and grass colored the slopes a sandy green, and evergreens supplemented their hues with life. Sheep started to appear, though not as much as they would near the bottom, and deer were plentiful along the slopes above the river. Mormon Ranch, the Flying B, and Bernard all displayed huge tracts of pasturable ground, and the open feeling of the country was relaxing. No doubt Andy regaled his guests during the leisurely portion of their trip with stories of the history that surrounded them. The Reberg Murder at Sheep Creek, Ma Crandall chasing off a robber at [what had become] the Flying B, Captain Bernard leading American soldiers through the backcountry during the Sheepeater Campaign of 1879; every mile on the river seemed to have its own story to tell. They ran a couple rapids the second day, but mostly they floated, fished, and took in the sights. They stayed on the river until they passed Waterfall Creek Rapids, where Big Creek came in from the left. Big Creek was the largest tributary feeding the Middle Fork, draining fifty miles of high country snow melt into the wilderness river. It had been the hunting grounds of 'Cougar' Dave, another icon of the backcountry life. Most importantly, Big Creek marked the area where the river environment changed from 'high desert' to a canyonland. Thereafter, they entered the Impassable Canyon.

The last day on the water was the roughest; the whitewater on the lower half of the river served as the perfect climax for the guests' river experience, and it was not unusual for everyone to get wet while crashing through Redside and Hancock Rapids near the end of the trip. Between the whitewater and the towering canyon walls (which stretched more than a half mile high above the river), the guests were astounded by some of the best views of the week. Everything prior took second place to the forever-moments experienced in the Impassable Canyon. Like everywhere else on the river, the water was crystal clear; it was easy to see the boulders that fell from the towering canyon walls in situ on the bottom of the river. Talus piled up for thousands of feet, landslides taller than the trees surrounding them. Eagles flew low down the canyon, and bighorn sheep stepped lightly on the ledges that lined the vertical walls. The river had more water in it than at any point further up, and the canyon was narrower. And as the features above ground changed, so did the features of the river. One could not help notice the surreal reminders of the river's power. High water barreling through the canyon in the late months of spring jammed logs and whole trees up onto boulders where they remained when the water line fell. Sometimes trees were piled like cord wood along the banks of a turn in the river, and occasionally there was a tree perched horizontally on a rock several feet above the water. The river was faster, the waves were bigger, and there were more rapids to negotiate. In one 7-mile stretch, there were nine of the biggest and most challenging rapids on the Middle Fork, including Weber Rapids (where DuBois had his only wreck in 1942), Rubber Rapids (which he called 'The Beast at the Bottom of the River), Hancock Rapids (which gave Andy nightmares), Devil's Tooth, and House of Rocks. Every rapid was different. Like any good show, the excitement built to a grand climax in the Impassable Canyon. And when they passed safely through the tall waves and rolling whitewater, Andy made a tradition of pulling off to the side and climbing up to a 'secret cave' so that his guests could sign the register for the 'Middle Fork Whitewater Club'. No one was excluded; if they made the trip, they were part of the club. It was great salesmanship and quintessential Andy Anderson. Then they made their way to the river's confluence with the Main Salmon, where they would load up and head for home.

Week after week, year after year, the Anderson's worked to perfect their craft and promote recreation on the Middle Fork. Incremental changes were implemented to improve the customer experience and, judging by the feedback received from their guests, they did a fine job delivering the whitewater wilderness experience. Every now and then, a 'freebie' would take a week of revenue from the summer, but Andy always felt that the resulting publicity was worth the economic hit. Besides, he felt fortunate to be able to make a living doing the thing he liked best, and as long as it provided for his family's needs, he felt he was successful.

Joe left the business in 1951, and Andy carried on without him. It was tough because Joe had been Andy's partner since the late '30s, even after returning from England with his new bride following service in World War II. Joe was a good-looking man with a winning personality, and knew how to complement his brother's efforts without a word being spoken between them. Almost every publicity piece mentioned Joe, and few trips to travel shows or the river were made without him. Andy trained a dozen boatmen through the years and mentored many professional river runners, but no one was able to replace Joe's energy or expertise.

It was during the early 1950s that Ted began to grow up and out of full-time involvement at the ranch. He enlisted in the Army in 1948. During a furlough later that year, he married Phyllis Ebberts in a ceremony in Elko, Nevada. He continued to make trips with his father and took leave to film a documentary on the river in 1949, but Ted had spent his youth in the Idaho backcountry and felt it was time to start his own life. He stayed at the Bar-X Ranch for several years while he worked at local mines, but toward the end of the '50s, he moved back to town and left the outfit for good.

Andy continued with his outfitting business. There was no angst in the family; Joe and Ted left for good reasons and Andy bore them no ill will. But he was a long way from done on the river, and his passion for the backcountry was strong. Still in his 40s, he had plenty of energy left for the work to which he dedicated his life. He invited other family members to work at the Bar-X during the busy summer months, and the family responded with enthusiasm.

Mike Anderson, Ted's son, said he took several trips when he was young and shared several family stories from those days. He made his first trip in 1960 when he was 10 years old. He helped get the gear together; it was heavy and the boat ramps weren't nice like they are today. A film crew making a documentary was floating with Grandpa Andy and they had two boys around Mike's age, so Mike made the trip to hang out with the boys.

"They were Italians," Mike said, "and they were always singing 'Volare'. I was in Dad's boat seated right behind him for the entire trip. They had Near Beer and let me drink some! They fished a lot, and there was a side stream they would put the fish in and film them, pretending they were salmon. That was supposedly what the documentary was about."

Mike helped in the kitchen at camp. He peeled potatoes and helped with dishes. He remembers stopping at Elk Bar and climbing up the cliff to the flat spot where he found several elk sheds. He also remembers flying on the Fairchild cargo plane, coming in late in the dark with the engines blowing flames, and his mom being scared silly.

Mike had lots of great childhood memories from the Bar-X Vacation Ranch.

In 1952, Andy stopped packing customers into the river and began flying guests and equipment into the backcountry. He used local flying services to transport them to Indian Creek, a popular put-in later in the summer when the water level was low. It had a large landing strip and a wooden boat ramp that made it a lot easier to get boats into the water. In August, when water levels were at their lowest, Andy would fly customers into the Flying B and pick them up there for a shorter trip down the river. Although they generated less revenue, the 3-day trips from 'The B' extended the season and allowed Andy to fully utilize the long, warm summer days.

In the late 60s, Andy sold the Bar-X Vacation Ranch property near Meyers Cove and moved his family and his business to Challis. A businessman from west of Fort Worth, Texas named Worth Boogeman and his business partner purchased the property on Silver Creek and renamed it B-C (for Boogeman & Carlisle). It was a good deal for both parties; Worth was looking for a place in the northern Rockies to use as a vacation home, and Andy wanted to be closer to his grandchildren. Challis was only an hour from Stanley, and he could manage his outfit effectively from there.

He retired from boating in the early 1970s, and sold his river outfit to Eldon Handy. He stayed in Challis to be with his family and mine in the surrounding area. He continued to work with the industry, and was a Charter Member of the Idaho Outfitters and Guides Association. He continued to prospect in the Yankee Fork area until his death December 21st, 1994 at 84 years of age.

Terry Remembers

(..archived recollection of Terry Anderson, younger brother to Ted, who lived at the Bar-X Vacation Ranch with his family and later worked with the Forest Service)

"Growing up on the Bar-X ranch until I turned about 8 or 9 years old, I had very little to do with the pack trips and river trips my dad was doing. I was just a free little kid roaming the mountains with my pet deer, dog (with a bell on so Mom could listen and know where I was), and a rooster. I did have a tricycle but could not ride it very far because of the gravel roads and trails. I used to fish with a willow and a Prince Albert can full of worms or grasshoppers. It was a life experience I would not trade for anything.

When we moved to Challis and as I started getting older and bigger, Dad would ask me to fly in with the equipment and help rig the rafts and then catch the last flight out. It was exciting to be able to fly in and out of the back country. Then when I was about 16 my dad asked me to run a raft with mostly supplies and one guest. It was a great education learning how to run the raft and read the water, help set up camp and clean up after the meals. One thing I remember was how hot the water was when doing the dishes, Dad did not want anyone getting sick so keeping things sanitized to the best of our abilities was important. Watching my dad deal with guests, entertain them, and see to it that they had a good experience on the river helped me later on in life, too. It helped me deal with people both in the military and in my work life. Of course, Dad's story-telling ability was par to any comedian by today's standards. I would watch him keep his guests hanging on to every word he was telling them, and then get that twinkle in his eye and drop the punch line. He was a master at it, I can remember my kids asking after one of his 'Anderson Windys' if that was the truth, and he would just tell them "well I've told it before".

Our equipment was not nearly as modern as the camp equipment of today. We cooked over open fires, very few outhouses, and the pack it in, pack it out was not in effect like today. We had to bury our garbage in those days. We did not have the good river bags of today, so we spent a lot of time drying out gear after a day of running whitewater. Our rafts were the old Navy recon rafts, homemade equipment and heavy compared to today's standards.

When I turned 18, I went to work for the Forest Service and volunteered to be on a fire lookout for the summer. I was given Sleeping Deer Mountain which had not been used since the end of the WWII. So, a good friend of mine, Ernie Cobbley and I started to work on the lookout and get it in operational condition again. That was a great experience in itself, we had to backpack a lot of the tools and food into the site, and build a spot for a helicopter to land on so they could bring in lumber and move the material over from Woodtick lookout which had been closed down. It was either that year or the next that I was approached by the Forest Service to take my boss and three bigshots from Ogden down the Middle Fork to survey the campsites. Since they knew I ran the river with my dad and had experience, I could take them down. We would stay at the Forest Service Guard Stations, so we did not need a lot of supplies.

I volunteered with a little feeling of nervousness. I had equipment flown in from Challis and they transported me down by helicopter from the lookout along with my companion a brown dog. We started from Indian Creek and everything was going along great until the last day. Their plan was to fly out at Bernard Creek and a camp cleanup crew was going to finish out the trip. I had been doing pretty well, I knew they were a little nervous to have this 19 year old kid taking them down the river but they were pretty naive about the dangers of a river and what lay ahead.

When we approached Tappan Falls on their last day, I pulled over to survey the falls just like my brother Ted and Dad had taught me. I had never run a raft through Tappan Falls, but I had asked my dad and Ted for advice; they said stay to the right and I should be fine. So, I dropped one passenger on the left side of the river to take pictures and I got set up with the other fellows. We donned our Navy life jackets that Dad had purchased. The life jackets were very large and cumbersome to wear, the kind you see in the WWII movies. I was determined to stay way to the right. Unfortunately, I impaled the boat on a large rock and water started filling the raft. It sucked us down to the point where I could not spin the boat loose. We were very close to the bank so I finally decided I could jump to the bank and take the tie down rope and maybe swing the boat around. First, I thought, I would throw my poor dog to the bank. She did not quite make it and down through the falls she went, swimming as hard as she could go. Next, I took a big jump to the bank and I also headed down the falls, only I forgot to let go of the rope. I hit the end of the rope so hard it swung the boat around and now the canvas bow was upriver stopping some of the water from filling the boat. Of course, I let go of the rope and pulled myself up the bank about 100 yards down river.

About that time, it started to sleet - being the spring of the year. I don't ever remember being so cold. Anyway, I started back up to the boat, they were all still sitting in the boat, I remember one of the Supervisors life jackets was down around his knees. I thought if he goes in, they will find him in Riggins upside down. He had also tried to take one of the oars and jammed it into the rocks or water and broken it in half. Somehow the paddle part flew into the boat and he lost the handle part. Being 19 and not thinking, I did not have spare oars brought in with the boat. Anyway, I helped them get to shore and we proceeded to pull on the rope until the boat finally came loose. By then it weighed so much, it just pulled the rope right through our hands and down the river went the boat.

By then I just wanted to cry. I saw my job going away and some kind of reprimand coming down the pike. They took off running down the trail and the boat drifted into the bank not too far from where I had climbed out. I started looking for a sapling to cut down and wire to the paddle for a makeshift oar. With some old Forest Service telephone wire which had been strung all over the country and maybe still is, and a sapling, I built an oar while they bailed out the boat. Someone had a bottle of brown water and gave me a swig to help with the cold. I think we did build a little fire to get warmed up. Needless to say, I was a little shook up. The ironic thing is the guy across the river got so excited he did not take even one picture. By the way my poor, wet, cold dog was fine, but not so sure she was into this new experience. I'm sure she was thinking why didn't he just leave me up on the lookout.

Anyway, we boarded the boat and proceeded down the river with one good wing and one half broken one. I had very little power with my makeshift oar, so we got into trouble on Lower Tappan Falls and had to bail the boat out again. By now these guys I'm sure were wondering what they had gotten themselves into, along with myself. I'm not sure how much farther down the river it was but we pulled into the bank to relax and fish a little bit. They got into a run of Steelhead coming up the river and I think they all limited out.

Needless to say, they forgot the wreck we had just had and all they could talk about the rest of the day was the terrific fishing.

At Barnard Creek, I had a couple of oars flown in along with the guys that came in to help clean camp sites the rest of the way down the river. They also were very naïve about the river and the dangers especially with some kid from one of the lookouts. We started cleaning up campsites and filled 2 or 3 large canvas bags in just a few camps. All the buried garbage, cans, bottles, pull tabs etc. were everywhere. I began cussing dad and all the other outfitters about burying all the garbage in the past. High water would come in and uncover the little garbage pits they all used so starting a river management system became very necessary.

Learning how to run the river proved to be a very beneficial trade with the company I eventually went to work for. The Truss Joist Corporation just happened to employ a lot of river rats so along with them and my family trips through the years, I have enjoyed floating this beautiful river. It has been and continues to be a great legacy for our family."

Salmon-Challis

I realized, while winding through the lazy canyon cut by the Salmon through the mountains, though decades passed since Andy left, not much of importance had changed. The landscape remained essentially the same and, for the most part, so did the culture in Central Idaho. When I started work at the B-C (on the same site using some of the same buildings as the Anderson's Bar-X Vacation Ranch), I lived in the presence of the recent past. The wilderness was protected and, for all practical purposes, change was officially prohibited. We still used stock to transport food, gear, and hunters to backcountry camps. The saddles were leather and the men were still tough, and the work was still long and hard. And the magic of the wilderness was still captivating for those of us instinctively drawn to it, and no amount of Progress or Technology was ever going to change that fact. The trails were still difficult and dangerous, and the river remained wild and free. The nation changed since Andy left the Bar-X Vacation Ranch, but his beloved backcountry was pretty much the same.

I drove south toward Challis on Highway 93, a two-lane blacktop that began in Montana at the Canadian border and rolled all the way down to Arizona. It was overlapped by the Salmon River Scenic Byway from Salmon to Challis (which then went west to Stanley with Idaho 75). I was leaving Salmon to meet with Ted Anderson for lunch in Challis.

'Salmon City', as the locals called it. The biggest town for several hours in any direction and home to approximately 3000 citizens, it was the biggest little town in Central Idaho. As such, it had a lot of the things people wanted when they came out of the backcountry. There were restaurants and bars, a bowling alley, and a small walk-in movie theatre. There was a high school, a couple hardware stores, and a good-sized supermarket with a pharmacy and a deli. But the closest Walmart was 150 miles away. With the exception of a mini-Burger King hidden in a convenience store, there were no fast food franchises, either. Home Depot, Costco, McDonalds, and such were a long drive across the Arco flats (a place so desolate that the government built nuclear power testing facilities there) or a trip north to Missoula, Montana. There was no available bus line, and no car rental; you had to drive or hitch a ride to the airport (which, for commercial flights, was minimum three hours in any direction). Businesses shut down pretty early each night, and everything was closed on Sunday. Still, Salmon seemed relatively connected to the outside world and served as a valued supply depot, and that was enough to make it a city.

There was a barber shop in Salmon; an old-school establishment with a red and white barber pole outside and 'BARBER' in big frontier letters painted on the glass street front. A bell hanging over the front door alerted the occupants of every visitor's entry. There were stacks of hunting magazines, Popular Mechanics, and newspapers on a little table to keep you busy if you had to wait, which you did sometimes because there was only one barber in town. There were a half dozen waiting chairs lined up along the mirrored wall on the right. On the left, there were two barber chairs; he used the one closest to the door, the other chair was a spare. Wearing a white smock, he stood in front of basins and shelves covered with museum-worthy man clutter; shaving soap in ceramic mugs, hair tonics, pomades, and combs in a jar of blue mystery liquid. Coffee cups with patrons' names on them dangled on a wooden pegboard by the vintage Mr. Coffee machine. The barber might be cutting hair, or he might be playing cards; either way, you might have to wait until he finished.

Several of us shared the opinion that our barber only knew one kind of haircut; you were free to request any style at all, and the cut menu offered tapers, blocks, and burrs… but you were probably going to leave with your hair trimmed a little on top, shorter on the sides, and blocked all the way around. And every time, he finished with warm lather and a razor shave on the back of your neck. He could hack out a burr, but he'd butcher a fade; he was a barber, not a stylist, and that's what you got for twelve bucks.

I have walked in to see the card table full but no barber in the shop. It was open, but not for business. The old guys at the table told me so, then went back to playing cards. Sixty years old and I felt like a kid in there, and smiled like a kid at the nostalgia. I had not been offended, just informed: I would have to wait for my haircut until the barber came back. It might be tomorrow, maybe the next day. I could call or come by later. Challis didn't even have a barber; when Ziggy died, no one replaced him. Unless you wanted to schedule an appointment at the salon, you had to drive to Salmon for a haircut.

When I lived in the Idaho backcountry, it seemed too far to travel without having a couple good reasons for going to town; from the Bar-X Ranch by Meyers Cove, it took at least two hours if the roads were good, and they were hard miles on the truck and tires. Sometimes it was impassable in winter. Considerable time and expense were involved in the trip, and it was best if I made a day of it. Typically it was a full-blown grocery run with stops at Murdoch's and the liquor store. I always went into Arfmann's sporting goods to see if they had anything new, and I usually walked the length of the mile-long main street. It was a clean, polite town lined with a few old two- and three-story brick buildings, lots of them looking a hundred years old. Some of the backcountry hands and cowboys looked a little rough, but it was a safe and hospitable place, representative of the American West. There was a nice little park built where the river ran through town, and a bronze grizzly bear hunkered over Main. I usually finished with a long late lunch at Brewster's or the Junkyard Bistro, then a stop for fuel on the way out of town.

In late spring and summer, you would see river folk in Salmon. Outfitters and tourists passed through town on their way back to the customary embarkation point. Almost all Middle Fork trips began at Boundary Creek, outside of Stanley, Idaho. After a hundred miles of whitewater and wilderness, they ended a few minutes past the river's confluence with the main Salmon at a place called Cache Bar. Vehicles were waiting at the end of the trip to take everyone back to Stanley. The outfitter's leaders said goodbye as they loaded the guests onto buses for the long trip back to hotels and cars, and the crew loaded all the outfitter's gear on trailers. Bright-colored 7-man inflatables were stacked three- to five-high and strapped down for the rough ride home, as well as tents, chairs, coolers, kitchen gear, and all the other equipment they would need to outfit their next group of guests arriving in Stanley. From the end of the trip on Cache Bar to Shoup was probably twenty miles and another 18 to North Fork, and the road was a torn up, barely-two-lane track that appropriately completed the wilderness experience. It seemed a long, bumpy ride just to get to the main road. At North Fork Village, there was a gas station and a convenience store, and US 93. Twenty more miles to Salmon, sixty more to Challis, and another sixty to Stanley. With a one-day turnaround between guided tours, the crew had no time to rest. The equipment had to be hauled back to Stanley and reprovisioned for the next eight-day trip. Since Salmon had a big store and laundromat, part of the team stopped to do laundry for the crew and pick up personal supplies. Others returned to Stanley to load groceries and meet the next complement of customers. Genuinely happy and remarkably hard-working, professional river guides exemplified the spirit of the Middle Fork. They were seemingly in love with the wilderness, and did not mind being with it every morning and night as long as the river allowed. Their enthusiasm was contagious, and added to the experience of a Middle Fork river trip.

The Salmon River rumbled alongside US 93 for the entire stretch from Salmon to Challis; it was a scenic drive best taken slowly. Bald eagles watching the water from trees at the river's edge were not an uncommon sight. Pickups pulled off into ad-hoc parking scraped into wide spots on the road, and they marked where locals pulled off to launch a boat or just fish at a good spot on the river. Crew buses with stacked trailers pulled over, too, and let quicker traffic pass. I passed a crew bus on my way down the meandering highway. I wasn't in a hurry; a young bearded guide waved and pulled over on a straight stretch in the road. And I waved back, to keep that friendly vibe alive. Never crowded but ever busy, there was a subtle energy in the summer when most everyone was working at jobs they rightly loved.

The mountains that surrounded the River of No Return (so named by Lewis & Clark's 'Corps of Discovery' when they rejected it as a possible waterway for their transcontinental crossing) were of several types and shapes. A lot of them rolled away from the road, soft in color and contour, inviting in a way. But in many places, the ground fell away and left sheer cliffs standing over steep and rocky slopes. Not particularly tall mountains, they were grand in scale. Sage-colored slopes were the backdrop for a high desert with short, steep mountains that broke into rust-colored boulders and coal-colored scree. Uninterrupted ridgelines stretched for miles, separating valleys like saw-toothed rock fences a mile high. Serrated ridges that marked breaks along a fault line; ragged compound fractures of the earth.

In contrast, steep slopes stroked and smoothed by the river for eons exposed a more subtle visual history. Layers of volcanic sediment, pastel reds and yellows and creams, marked clearly the durations of separate geologic events, especially on the north end of the round valley of Challis. So much of planet's history was illustrated in lines of color across the naked hills that rose above the river. Central Idaho was the western shoulder of Yellowstone and geologically linked to its geothermal potential. Earthquakes in the area were testament to the character of the region.

Metamorphic like most of Earth's crust, the base layer had been folded up with ancient sea beds. Tectonic shifts pressed the sea floor between masses, and all that pressure created heat. Magma squeezed out along the continental edge to form the deep rock Idaho Batholith. Things folded and faulted and boiled for an unimaginable length of time until the whole thing exploded and formed the Challis volcanic field, easily identified by its enormous calderas and massive granite outcroppings of crystallized magma jutting like broken bones from the ground.

"By this time Future Idaho was starting to look like a battleground, with old volcanoes scattered around, fault-block mountain ranges forming and eroding, rivers that were cutting deep canyons, and giant, but temporary, lakes dotting the landscape." (Bill Bonnichsen, Geologist, University of Idaho)

Forged of intense heat and pressure, the landscape was beautifully hostile, not for the timid. The facade of forests and rivers and wildlife disguised the fact that Central Idaho remained geologically active. The massive earthquakes in 1983 on Mount Borah, Idaho's highest peak, indicated the mountains were still rising along a fault that ran right up along Round Valley. The region had been hot and violently alive for as long as Man had measured Time, and there was no reason to believe it would quiet down any time in the future.

In spite of its violent backstory, Central Idaho was peacefully green on that day in late May when I drove along the road beside the river. Enough rain had fallen that week to feed a spring bloom and swaths of yellow flowers colored the pastures. The sky was blue, though hazy with the smoke from something in California. A coyote trotted through a hay field being watered by an irrigation pivot, and cattle watched him as he made his way along a wheel track made by the machine. Farmers cultivated everything they could turn into a field, as well as the slopes that surrounded the fields; they made use of every acre of fertile soil. Farmers or ranchers who were growing their own hay populated the good land along the river and in the valleys between the mountains.

There were a couple little villages along the highway between Salmon and Challis; some were just a gathering of homes. I curled around one bend in the river that leaned left so long, it almost made a circle. And all around the edge of the road were little houses and mobile homes, as if Salmon or Challis was too crowded for them. Four-wheelers and snowmobiles were parked next to sheds too full for any more machines, in yards that overlapped, and everyone had a grill. Not tacky – just country… populated by real people who would stop to help anybody in a bind. It was not unlike the Texas of my youth; there were cultural similarities. Towns between towns on long lonesome highways, and stores run by people that lived in the back. Diners that might serve a dozen meals a day, where teasing and tipping was encouraged, and respect appreciated. A little humility paid dividends in that part of the country. If someone waved at you when they passed, it was not a mistake; just courtesy.

The river was most alive in the spring. Snowmelt from the

mountains raised the speed and depth of it, the surface rippled, and small birds darted angles across it looking for tasty treats. Cliffs rose abruptly, color-coordinated with earth tones delicately matched; chocolate brown splashed with rusty orange, patches of gray with green lichen shade. Tufts of grass where soil accumulated enough to allow its growth. Along the shore, cottonwoods reached up with bright green brushstrokes that swept up from their pale white bark, and they marked the waterway's path from a distance. And if you stopped, if you pulled off the highway almost anywhere on the river side, you could hear the water... the sound of the river moving through the landscape, cutting on the canyons, shaping the world.

The countryside permeated the lives of the people that lived there. You didn't have to live off the land, but you certainly had to live with it. The way people dressed, the vehicles they drove, the way their houses were built and utilities organized; so many aspects of life taken for granted in 'the civilized world' just did not work around there. If you lived or worked in the backcountry, you needed tire chains in your truck, and you needed a chain saw, bar oil, and fuel, as well. It was wise to keep a couple gallons of drinking water, an extra coat, and rain gear. They were not optional; sooner or later, you would need every item. And that was just for the truck. There were good reasons for a lot of the lifestyle choices made by its inhabitants, and only a few of them were cultural. Mostly, people had to adapt to life in Central Idaho if they liked it well enough to stay.

Of course, there were those that lived solely occupied with the Business-of-the-Day (tethered to television and smartphones and all the digital stuff), detached from and oblivious to the different ranges of Rocky Mountains that surrounded them. Unfamiliar with the Beaverheads, the Bitterroots, the Bighorn Crags, or Lost River Range. Never setting foot outside their comfort zone; never venturing into the wilderness so abundant in Idaho. They were happy enough with looking at life while going through the motions of living it.

But most of the people in the central part of the state, especially those in small towns, were intimately acquainted with their geographical surroundings and all their natural resources. Everyone had favorite activities, and many activities were seasonal. And almost all of them took place outdoors. There seemed to be something specific to every season of the year.

In the winter, of course, weather dictated what could be done. For most people, it was high school basketball and wrestling, community events that could be organized indoors. Winter was a quiet time for businesses and individuals involved in tourism. Winter tourism was pretty thin in Idaho, with only a handful of goods ski towns (Sun Valley the exception). River outfitters were in hibernation and hunting outfitters hunted mountain lion, which was too rigorous for most out-of-state guests. As nature suggested with longer nights and shorter days, winter was the time of year for rest.

In the spring, steelhead fishermen crowded every bend in the river. Restaurants and hotels got their first taste of tourism as anglers came from everywhere to fish for steelhead in the Salmon River. They were fish that began life as rainbow trout and then migrated to the ocean. They were considered by many to be tastier fish than salmon, and healthier, as well. Steelhead attracted anglers from all over the West, each competing with the others for the best fishing spots. Even before it looked like spring, while there was still ice on the river, guys nobody recognized showed up for breakfast in local diners. Way too perky for that early in the day, they'd strike up a strained conversation with a regular in a room full of regulars, and ask about the best place to fish.

As spring progressed and the weather warmed, ice and snow melted off the mountains, and water in the creeks and rivers got deeper and wider and faster. As the run-off increased, bold boaters waited for 'high water' to provide them with maximum speed and flow, and braved the swollen waterways as a kick-off to the season. It marked the beginning of the annual lifecycle, and all the plants and animals and people began to bloom at once. Immediately and instinctively, everyone was busy making full use of the three seasons that followed. It was time to make repairs and fish a little bit, to put up wood for the next winter if you were industrious. Time to get the garden ready and start some seedlings indoors, to open up the summer home and move what was needed there. Time to put inventory up for the tourists that were coming, to order new products, and to get employees back on board in time for the summer season. It was time for the Forest Service to open campgrounds for the public, and to clear trail of rock fall and dead trees. For farmers to burn grass from their irrigation ditches, and farriers to shoe stock. Winter was always long and hard, and spring was always welcomed.

In the summer, it was all about tourism. A lot of people will tell you that mining and manufacturing were the biggest contributors to state revenues, but the Idaho Chamber of Commerce said in 2017 that tourists spent $3.7 billion in the state that year, making tourism the third largest industry in Idaho (behind agriculture and technology). And tourism in Idaho happened from May to September; through November, if you included hunting season. In the summer, the river was the main attraction. Approximately ten thousand visitors each year rafted down the Middle Fork, and everyone from outfitters and guides to hoteliers and restaurants benefited from the seasonal activity. Everyone was busy because business was good, and it was growing year-over-year. College kids out for summer spent every week guiding on the river or working at a resort, stacking cash away for the school year. Restaurants that would be closed all winter used every available hour to cover in a few months their costs for the entire year. Hunting outfitters were clearing trail and stocking camps. Businesses not directly related to tourism benefitted from the summer activity. General retail was up, as was consumption of commodities such as food and gas. It was the best time of year for new construction, and there was always something to remodel or expand, so contractors and construction crews used the long days of summer and made haste while the sun shined.

In the fall, big game and rough country attracted hunters from all over the United States and abroad. There were no less than 250 licensed hunting outfitters and related services registered in the state, and from mid-September until the week before Thanksgiving, the mountains were full of men on horseback looking for bear, big elk and mule deer. Fathers and sons, husbands and wives, and generations of family gathered for hunts. Some were for trophies; all were for meat. In Idaho, athletic hunters could pursue difficult prey. Bighorn sheep and mountain goats could be found deep in the Idaho backcountry, and only a lucky few were granted the permits required for the controlled hunts. They were physical affairs that tested the hunter's strength and stamina. Hunting was a cultural link to everyone's ancestors, and it was an important part of life in the region. Even after Thanksgiving, those willing to hunt with a bow were allowed another month to pursue meat for the winter. By the end of December, though, most people retired to the comfort of a home well stocked with wood and food to rest for three or four months until spring came back around.

I had just emerged from a long winter in the Idaho backcountry. I appreciated the change in season and was energized, like everyone else. I felt it as I drove down the highway. I was amused by every little thing, and smiled at every opportunity; I felt the synergy of life as I drove through it. And I knew that almost everyone around me felt it, too.

Nine miles north of Challis, I passed Morgan Creek Road. Forest Service Road 55 was the turn-off to the backcountry, and the road the Andersons used to get to the Bar-X Vacation Ranch starting back in the 1930s. Custer County (Challis was the county seat) maintained the road up to Morgan Summit, but the other side – the Panther Creek side – was unclaimed by either county (at least when it came to maintenance), and the rocky dirt road was pitted and potholed, and it rattled with washboard ruts. Panther Creek ran all the way to the Salmon River, joining a few miles upriver from its confluence with the Middle Fork. There was a lot of history on that road, and a couple of towns that were there in the '30s had completely disappeared. I imagined the road every bit as rugged and remote as it was in Andy's day. The scenery was probably different in a place or two, but the way into the backcountry remained the same.

Past Morgan Creek, US 93 felt like it flattened out a little as it swept more broadly toward the valley. Through big pastures that spread all the way from the river across to the slopes on the other side, the road curved up and right, then bent left toward town. It was higher above the river, which was on the left, and the view opened up as I entered the north end of Round Valley.

I read somewhere that Bourdon's brigade of Hudson's Bay Company trappers discovered Round Valley in 1822, and continued to work the area for a decade or more. Prospectors arrived and in 1870, miners discovered gold on the Yankee Fork, that same period of time that mines started popping up everywhere in Idaho. Not long after, A. P. Challis saw an opportunity and seized it.

Challis, who had been in John Stanley's gold discovery party through Stanley Basin in 1863, set his eyes on Round Valley and organized a settlement. He surveyed a town and started a ranch, then built the town to support the mines in the mountains that surrounded it. Some of Challis's historic resources were preserved in the mining and supply town architecture, the way that several stone buildings were constructed from rock recovered from the bluff above town. Not long after A.P. and his friends got settled, the town grew dramatically. According to the census of 1880, it had developed to include three hotels, four saloons, and several new permanent dwellings. There were local sawmills at work. A lithograph from that period, though not absolutely accurate, displayed a four block-long Main Street of one- and two-story false front or gable-front facades. For two blocks north and south of Main, the streets were lined with rail fences that enclosed small houses on generous lots.

Many immigrated to the region to find their fortunes mining in the years following the Civil War, and the town named after A. P. Challis profited from mining and the businesses that supported it. When the railroad completed a line to Blackfoot in 1879, a Challis-Blackfoot stage line began regular service, and a toll road was cut into the Yankee Fork from Challis.

The settlers who had come to Challis were predominantly Mid-Atlantic and Midwestern, but a substantial number were immigrants from England, Ireland, and Western Europe. Men outnumbered women five-to-one, and there were few children listed in the census. Slightly less than half of the resident population were miners; the remainder were occupied with services and trades necessary for the subsistence of Challis and the surrounding mining towns. Many people were occupied as laborers, teamsters, freighters, and packers.

In 1880, Challis had thirty-three carpenters and two stone masons, a substantial number of builders for a town of 614. The area's eventual change from a mining economy to an agricultural one was anticipated by the town's cattle rancher, ten sheepherders, and twelve farmers. When Custer County was carved out of the existing Lemhi and Alturas counties in 1881, Challis became the county seat. Between 1888 and 1896 a road along the Salmon River (following the present route of U.S. Highway 93) from Challis to Clayton and

Sunbeam was built, and the Challis toll road became a free route. The town survived as a pack trail station that linked Custer, Salmon, and other mining towns with the Utah and Northern railway.

> *"The citizens of the town are hopeful of the return of the good old times of the prosperous mining days gone by. Mines of rich ore are still nearby, and when work is resumed Challis will be the busy, bustling town as of the past."*
> - The Silver Messenger, 18 September 1894 [p. 1, col. 4]

By 1890, with the gold rush over, the town lost a measurable part of its population. To depress things further, a fire in the commercial district in 1894 destroyed a number of businesses. There was a burst of enthusiasm when the Yankee Fork of the Salmon River showed promise. But by 1900, the population of Challis had fallen to 387. A few years later, the Yankee Fork mines were mostly shut down and the Round Valley economy was forced to finally make the change from relying on mining to acknowledging the importance and stability of a ranching and agricultural base.

Through the first three decades of the twentieth century, Challis's population stabilized and steadily grew to 803 people in the 1930 census. There were few new business additions to the original townsite. In general, it remained a humble town whose buildings were small in scale and modest in design. Main Street, the wide thoroughfare that divided the town north and south, was mostly a one- to one-and-a-half-story streetscape with a Rocky Mountain background. The commercial buildings included dressed rock structures from the early 1880s, and false- or gable-fronted frame and log buildings from as early as 1895, randomly mixed with a variety of older structures covered with diagonal siding or board-and-batten, and newer structures in cinder block or clapboard… all lined up on Main Street presenting fronts of diverse textures and designs. The craftmanship of the dressed rock facades represented skilled work.

The residential architecture of Challis was also modest in size and, in most cases, design. Probably the oldest existing houses, some built of logs, were south of Main Street. North of Main, at the foot of the bluff, were the more elaborate two story dwellings. Except where newer houses or trailers were added to subdivided lots, Challis homes stood on relatively large lots set back several feet from dirt streets behind cottonwood, locust, and silver maple trees.

The layout of Challis was common to Western mining towns. It was a supply center more than a mining camp, so it had only one commercial street until the recent development of businesses along U.S. 93. Residential areas conformed to a grid parallel to Main Street except in the northern half of town, where the creek bed and the slope of the bluff forced streets and house locations to conform to the lay of the land. No public square or park was originally planned for the town, although newspapers from the 1890s mentioned areas for racing and picnicking.

Even though the town was located on a highway and near a railway terminal at Mackay, a degree of isolation persisted, as it seemed in all the Salmon River communities. While it was possible to travel from one small town to the next, it was not practical unless necessary. Everyone learned to live where they were, and adapted to changing times and seasons.

In the late 1930s, Andy Anderson moved into the area to begin pursuing his dreams. Although their mailing address was in Forney, Challis was the town Andy used to grow his outfitting business. Ted went to high school in Salmon City when he lived and worked at the Bar-X with Andy, but he moved to Challis when he got married and started a family of his own. It was the town Ted called home for his hitch with the Forest Service, and where he retired after continuing the family's tradition of service to the Middle Fork River. Challis was home to the Andersons; a most appropriate backdrop to their story.

Changing Times

Just past the airstrip, speed limit signs slowed things down to 45, then 35 mph in town. There was Gateway MiniMart and Kimble Oil; two filling stations on the right just as you came into town for easy gas-and-go. Challis had plenty of gas stations.

Across the street was the Forest Service setup. A good-sized house with a sign that said 'Supervisors Office', and smaller houses that served as residences. There were some maintenance buildings, fenced areas, a couple parking lots. Years ago, it was headquarters to fifty or sixty people, but it had been downsized substantially. The facility still housed the Challis Ranger District, but only a couple of people lived there in the summer, and it was closed during the winter. I had seen a lot of them in Idaho; fine government facilities that represented the priority our elected officials once put on responsibly maintaining our natural resources. Simple well-built structures, professional installations, all painted the same… noticeably official. So many had been moth-balled, left losing the battle with Time. Funding for these stations was lost over a period of time and, once lost, rarely recovered. I wondered how it looked when Ted worked there forty years before. But I laughed to myself, knowing the answer; it looked pretty much the same, except maybe the paint was fresh.

Which was true for Challis in general. It had not changed much in the last few dozen years. It was not a ghost town or a sad town, or run down from years of inactivity and neglect. Challis knew very well what it was; humble, understated and satisfied with the deal. 'The Deal' was being able to make a living in a place like Idaho. It wasn't perfect for everyone, but damn near perfect for some. And those people, like Andy and Ted, were glad for any stable work that allowed them to stay in the area. There was no easy way of life in that part of the west, and that was part of the culture; men and women alike in Idaho did and had always done whatever they needed to do to make a life for themselves and their families. A study of history revealed the families that built the central part of state; people who immigrated to hard country and settled in to stay. They are the family names of the ranchers and farmers who homesteaded and still lived and worked in Round Valley, Mackay, and the Pahsimeroi. They were proud of their history and happy with their proven way of life.

The town provided the basics for itself and the wider community, and a burst of summer traffic that pretty much stayed on US 93; there was fuel, food, and spare parts (for vehicles and homes). There was a bowling alley with a drive-thru window on the kitchen, and a handful of simple restaurants in town. Nothing franchised or particularly fast. Since it served as the county seat, it supported area telephone, internet, and electric service companies. There was a building supply and a Family Dollar, and a liquor store across from the bar. Those businesses catered to the general population and were able to stay profitable whether the mines were open or closed.

I turned right onto Main Street and drove past the city park. On the right was the Challis Roadhouse and a long 'L'-shaped one-story motel, wooden like the attached restaurant. Lunch at the Lodge featured a chalkboard special for a $5 burger and fries. The business changed hands a couple of times in the six years I'd been around, but the chalkboard never changed. I went there from time to time when I was in a hurry. But I drove past and parked my truck on the street outside the Y Inn.

It was called the Y Inn because, back in the day, the road came to a 'Y' intersection and the restaurant was built on that spot. It was two large plain log cabins connected by a breezeway with the sign fixed above a single-gabled roof. The sign was a tall rectangle of five plastic squares, alternating yellow and white, with a big red 'Y' standing on top of the rectangle. Below the 'Y' in matching red letters was 'INN' in the first square, and 'CAFÉ' spelled out in black block letters in the bottom four. Glass doors opened up to an empty lobby where a long hall to the right led to bathrooms, and the restaurant was through another set of glass doors on the left. A sign said "Seat Yourself", but I stood for a moment to look for Ted.

The log cabin décor continued inside with a dark lacquered wood ceiling. Four lighted ceiling fans kept the dining room cool. It was full of small square metal lunch tables with brown faux wood table tops. The chairs were framed with square tubular steel, black like the tables, and comfortably upholstered with black vinyl covers. The tables stood alone for groups of four or less, or were grouped together for six or more. And near the far corner in the open part of the room, there was a round table. The far wall was adorned with photographs of flora, cropped and mounted and displayed on the wall 'for sale'; the proud work of some local artist, part of the community's pride. Country and contemporary music leaked from a radio in the kitchen and blended with conversations in the dining room. There was a father and son, a couple guys on break, two old ladies, and two ol' gals… and one guy sitting alone at the big round table. They served tater tots and fry sauce, and pop came in translucent red plastic glasses from the Coca-Cola distributor. It was as American as a Norman Rockwell painting.

The personalities of the people in the Y-Inn were an enjoyable part of the setting. One lady sat upright, her left hand in her lap clutching her napkin while her right hand worked her fork. At another table, two members of a road crew hunkered over their meals, with hoodies pulled up over baseball caps and beards up to their eyes. An elderly couple tasted each other's fries, and enjoyed their free time together. The clock was over the 'Employees Only' sign above the kitchen door, and it showed straight-up twelve o'clock. I turned in time to see Ted come through the door.

He had just celebrated his ninety-first birthday. And there was nothing about Ted that suggested his age was any sort of problem. He walked upright with strength and balance, drove a shiny truck, and roamed through the dining room before he took a seat to catch up with all the other locals who made the Y-Inn their place. Some were hard-of-hearing, a handful had not aged as well, but Ted was like most of the other seniors in Challis – savvy, sometimes a little sore, but certainly not old.

He greeted me with a handshake and a smile. His face was bright. It was good to see him again.

"How are you, Ted? How'd the winter treat you?"

"Good. Great. Can't complain. Got this new knee, and its working out real fine."

"Does it hurt?"

"No. It's no big deal anymore, those surgeries. I was in there for two nights, then right into Physical Therapy. That's the hard part. But they opened up the golf course now and I'm ready to go play a few holes."

He was wearing blue jeans and sneakers, plaid flannel with a t-shirt underneath, and a red-and-white ball cap peaked the old fashioned way. The brim was curled like a baseball mitt wrapped around a ball to break it in, a throwback to the Greatest Generation. They seemed to be the first to wear ball caps in the military, and it was their style to wear them high on their heads. Ted had his cap parked up on his forehead like they did back in the day, so you could see his eyes.

They were eyes that had seen more in the backcountry than could be seen nowadays. Not just on the river, but on the trails and in the drainages of the wilderness surrounding it. Before the Frank Church Wilderness was created, when it was still the Idaho Primitive Area, Ted packed gear to remote ranches, leveled runways with Fresno plows, and irrigated to grow hay for the stock they needed to do their work. He was a teenager, and had to drive two hours from the Bar-X Vacation Ranch to Salmon when it was time for school. Deep Creek wasn't open back then, so he drove up and down Napias Creek, a treacherous mining road too steep to be safely driven in his brakeless truck, but the only way to town. Ted had been living and working in that particular section of the backcountry since he was a teenage boy, and he knew every ridge and natural landmark. His knowledge was embedded; he remembered things very well. His eyes sparked in conversation as he recalled the past.

"So how far were you up Camas Creek?" he asked as we settled in. Our last meeting was the week before I left to spend the winter alone in the Frank Church Wilderness.

"I spent most of my time at a base camp by Castle Creek. Right there at the confluence. I had a second camp established eight miles up at the South Fork."

"Up past Strickler's place?"

The Strickler family homesteaded a ranch in an area later 'grandfathered' into the Frank Church Wilderness. The property was exempt from wilderness restrictions, even though it was surrounded by protected habitat. It was located in a hidden valley six miles inside the south-eastern boundary of The Frank with steep canyons and high alpine lakes that formed a natural perimeter. It was a four-hour drive from Challis and (unlike most backcountry ranches) did not have an airstrip. I called to ask permission to cross the property during my excursions that winter, and passed through it several times.

"Yes, sir. A couple more miles or so past Hidden Valley Ranch."

"But you had Castle Creek Ranch just a few miles up that hill."

"Yeah. I could have just walked up there if I got in trouble."

Ted was familiar with the area where I spent the winter, a remote drainage that few people knew about and only a handful ever visited. He teased that I was only four miles from the nearest ranch, which he knew was way up hill and deserted that time of year. But he also knew I skied up there to shower and use the Wi-Fi twice a month. Back in Ted's time, there were still some trappers who spent the entire winter in the backcountry, moving from hut to hut harvesting fur and resetting traps, until spring came and the fur got thin. They didn't take breaks at their rich friends' ranches or check-in with family on the internet. They were tougher back in Ted's time. He knew that kind of life, had watched his dad live it, and knew from personal experience that some people couldn't get enough of that life. Many were those that braved a sample, but few were willing to venture into the heart of the experience and stay there. Thousands of people rafted the Middle Fork every summer and camped on its beaches and in its canyons, but the wilderness surrounding the river remained cloistered and secretive. I had seen a little, and knew enough to talk about it; that was my 'common ground' with Ted.

The waitress walked by a couple times taking care of other guests.

She knew Ted and probably everyone else in the room. I asked for a menu because most people came for the specials. They were written on a small whiteboard on the easel near the server's area in front of the kitchen. Of course, the regulars knew which specials were served on what days of the week, and planned their visits accordingly. On our day, it was

"Cheeseburger & Fries or Tots : Soup – Hamburger Macaroni"

The soup pot and ladle were right behind Ted, where waitstaff or customers could serve it. There were napkins rolled around soup spoons and stacks of bowls under the counter.

Turned sideways in his chair, he looked at the special, and I looked at the menu. The waitress came by and took my order for a patty melt sandwich. Ted said he wanted some soup.

"So what happened at the ranch in the 50s?" I fiddled with the stuff in front of me, moved my phone to the side, and lined up my silverware in preparation for the plate. "After the war, it seemed, business picked up. But Joe left, and then you left. Why did you leave the ranch when it seemed business was starting to get steady? Did you leave the Bar-X to join the Army?"

"I think so. Dad was still there. About that time, we were starting to fly customers in."

"To Indian Creek?"

"To wherever it was best, depending on the time of year. I came back from the service to make that one trip. I told you about that one."

It was the trip Andy organized with the Brooks from Boise, the one with Al Brick from Fox Movietone and Loomis Dean, the photographer from Life. Ted helped with that trip because Andy needed extra help, and it served as the foundation for several important magazine articles.

"What I had done, see, is I just got out of high school in '48, and I signed up for three years in the Army. I went to the service that fall. It wasn't so much to leave the ranch as it seemed like an obligation. I volunteered, which was because they were having the draft, ya' know. So I went ahead and signed up and went to Fort Ord. That's where I did my basic training. I came home on a furlough and that's when I got married to Phyllis."

"Was she from around here?"

"She was born in that log cabin there next to the post office. Her family was from Custer. Her and three brothers went to school here. She was several years older than me, so she was already out."

"So ya'll stayed around here…"

"Sure. She loved it out at the Bar-X, and everybody loved her. Big laugh and a great big smile. She could do almost anything."

"So you got time off to do that trip in '49, and then got out a year later…"

"Yeah, something like that. I wound up at Fort Warden in Washington, up there in Port Townsend. We went from there to Hawaii, and had maneuvers over there. I was a mechanic. I went to Mechanic School in Fort Ord. Motor pool stuff; the old LCMs [Landing Craft Mechanized]. We transported supplies from the ship to shore, and then to the supply depot. When we got back, the Army had decided they wanted to make it a career for me. But I was only making $75 a month as it was, and half of that went to my wife. $50 actually. She got $50 and I got the rest. That was the family allotment. It wasn't a career I wanted."

"So you got out."

"I got out. I served, but it was between the wars. You see, actually, I'm not a veteran."

"What do you mean? Of course you are."

"Well, you know. Not a combat veteran."

"So what. You got lucky."

"Yeah. World War II ended right when I went in, and the I got out before Korea. (laughed) I didn't complain at all."

He looked around again, distracted.

"I guess I get my own soup," he supposed out loud. He didn't mind, but he was ready to eat.

"This the guy writing about you?"

A tall guy brushed him as Ted stood up to serve himself. Clearly old friends, they shook hands and Ted started laughing.

"Frank! How do you know about that?" Ted asked.

"Oh, I been in on it. I heard about it from somebody over there," he said, pointing at the round table by the kitchen. Then he looked at me and winked, "Just put in there that he is the worst cribbage player in Custer County."

Ted shook his head and laughed.

"The worst," confirmed Frank. "There was no joy in beating him."

"We worked together when they first put the telephone line in," Ted added. "This is Frank Leaton. We used to play cards during our lunch break."

I stood up to meet Frank and to shake his hand. He was an easy-going guy, with loose clothes and new sneakers. He had a long wooden walking stick that I was sure he took everywhere.

Ted turned to the waitress, who happened to pass by right then.

"Do I get my own soup?"

"Absolutely, if that's what you want. And I'll bring you some ketchup and crackers, Ted. Do you want any ranch or fry sauce, sir?"

"No, ma'am," I answered. "No fry sauce. But I will take a little ranch dressing. Thanks."

Frank nodded 'goodbye' and moseyed out the door.

"I'll be right back with that," she said.

Ted turned to the soup tureen and loaded up one of those indestructible little beige cafeteria bowls with hamburger macaroni soup. He took the napkin off his silver and laid it on his leg, then waited for the waitress to return with saltine crackers.

"You don't like fry sauce?" he asked.

"I didn't even know what it was when I moved up here. And no, I don't like it much. I think it's an Idaho thing."

He chuckled, tasted his soup, then worked our conversation into lunch.

"We came back to Flourspar," he continued, "down there on Camas Creek. We lived right there at the mine."

"Back there by the trailhead? What was going on there? Was it a good mine?"

"It only lasted about four or five years. Jack Simplot had it. He bought it from a guy by the name of Chambers. We put out fluorspar that was 99% pure from that old mill. It looked like flour."

He paused to take another bite, and I dipped sweet potato fries in ranch dressing. It was perfect with my mushroom-and-swiss burger, a treat I waited for all winter.

"They used 'spar in metal making, chemical processes. Even use some of it for microscopes and telescopes. But back East, wherever they were doing a lot of farming, Kentucky and back in there, they picked it up out of their farmland. They had a lot of it back there."

"Did you live at the mine? Or drive back and forth?"

"We lived in those houses along the side of the road where the trail is now. The foreman lived in that house across the creek. The manager lived in a log home down there closer to the trailhead, further down the creek where Mickey is now."

"So what was it like living there? You lived there with your wife?"

"Oh yeah. Nice little camp. About 50 of us."

"Did they have a chow hall?"

"Yeah, for the bachelors that lived in the bunkhouse. We lived in one of those houses. My son turned one year old there."

"How many days a week were you at work? Did you go to town now and then?"

"Well, we worked 5 days a week. Eight hours a day. That was the style, ya' know. And we'd go to town, yeah. We'd even drive in for the movie sometimes, leave in the afternoon."

"So it was normal."

"Normal to us. What'd you think it was?"

"I got the impression if you worked in the mines back then, you worked your ass off all the time."

"Most jobs around here required that you work your ass off," he smiled. "My job at the mine was pretty normal. I worked some at the site, and did some mechanicing, too."

"What was Challis like back then?"

"It was a lot smaller. That was back before they built all these houses… before the mine went in at Thompson Creek. There was Main Street and a couple streets over that way, and maybe five streets over that way. Five hundred people or so. Might have been up to 600 by then. Clayton Silver was still going. Cobalt Mine was going."

"You ever work at Cobalt?"

"Oh yeah. I worked there a couple of years back in the '50s. There were a lot of people livin' back there. You know where Panther Creek Inn was? That was the last stop. We all stopped in there for a beer at the end of the day."

"Did it get rowdy?"

"Sometimes. But not me."

"Not you."

"Oh no," he grinned. Then he finished the last of his soup, wiped his mouth, and put the napkin in the empty bowl.

"Seems like a hard life to get used to…"

"Oh, it is a lifestyle, alright. We had guys come from Butte to work the mine, then when things slowed down, they'd go back to Montana. Move from place to place with the work. That was their life."

I could relate. When I was a young man, finally free from school, I followed the work wherever it was. It was normal for kids from my social class to labor at a variety of jobs before settling into a career. In Texas, it was construction or oilfield work. In Idaho, it was mining or something in the saddle. And there was the whole '50s dynamic at work; Ted was molded by a time when men served their countries and sacrificed for their families. Wives, too. Generally speaking, people put their desires to the side in order to give priority to the family.

I tried to imagine Ted younger, like a picture I saw hanging in his hall. He might have been in his fifties, sitting in a row with Phyllis and his kids, and a grandkid. It was taken during a family trip on the river. Ted was a big guy; not fat, but stocky like his dad. Heavily muscled and comfortably decked out in tank top and shorts. The shorts dated the picture from the '70s, but they might just have been old shorts. He had his cap on, propped up, and his head tilted away from the camera. His hair was dark, his skin tanned, and his mustache was thick and wide, with a smile tucked underneath it. Phyllis sat next to him, her eyes turned up by the grin on her face, her short blonde hair excited by the wind and the river. Mike and Tammy were a reflection of their mother's looks, the same eyes and smiles, and Mike was bent over hugging his son. Closer to the lens, Mike's wife was hugging her father-in-law. The whole family was having a great time. I remembered a comment that Mike made one time, that it was like living 'Leave It To Beaver'. It was a '50s sitcom that was typically wholesome, refreshing, and funny, and I could see that time in our country's cultural history reflected in his family picture. That picture was from the '80s, so it took young Ted a few years to find his calling.

"So in '57 you opened a bar?"

"Yeah, that's about the time Al Fiddler and I – we didn't open it. We didn't own it, either. We kinda rented the bar. Kerr's; it was over there where they had the tax place. Right next to Bux's there. Al and I met here in town. He was a musician and we all drank a lot, ya' know, so we got together. It was one of those things that happen."

"Was it fun?"

"Well, I don't know if you call it fun [laughing], I got to be my own best customer. That's when I went to work building those telephone lines. Arrow Construction was the name of the outfit. Then I went to work for Allied."

He thought about it for a minute, and reviewed the timeline following his departure from the family business at Bar-X Vacation Ranch.

"Flourspar had closed down, and then I went to work at Cobalt. It wasn't called Cobalt then; it was called Copper Creek. Worked at the mines until I did the bar thing with Al, and gave that up to work with the construction outfit putting up telephones. And after that, I went to work for Allied."

"What did you do there?"

"I was a handyman. Pulled lumber. Sold appliances."

"They sold appliances at the lumber yard?"

"Yeah, they had appliances and some furniture. I installed a lot of carpet. That's why my knees are bad. That's tough work, kicking that carpet in and stretching it. I was there for about 13 years. That was when we moved into that house on Main, when Tammy was one year old. Before that, we lived in a trailer up the street."

That was in the early '60s. Ted worked at Allied until '74, when a job opened up at the Forest Service that turned out to be a perfect fit.

River Manager

The Middle Fork of the Salmon River was exceptional fish habitat. Cold, clear waters from melting snow tumbled out of the Salmon River Mountains and into the boulder-strewn waterway, wild and remote, and federally protected. The last of the spawning salmon arrived in June after an incredible upstream swim. The big fish were finishing their courtship rituals, so the next year there would be a new generation to play their important part in the ecosystem.

The extreme migration of the salmon was one of the animal world's great journeys. Homing in on their birth stream, they fought strong currents for hundreds of miles, gained approximately 6,500 feet in elevation, and overcame physical barriers to return home to spawn. Salmon were particular spawners, not only returning to the stream where they were born, but also often to the same shallows. Their offspring emerged to make their own journey to the sea. Always a challenge, the migration had become more deadly in modern times.

According to Fish & Game records, native fish populations were in free-fall throughout the Columbia River basin, a situation so dire that many groups were urging the removal of four dams to keep the fish from being lost. What was once the most productive wild salmon habitat in the world was witnessing a dramatic loss of territory. Before the 20th century, some 10 - 16 million adult salmon and steelhead trout were thought to have returned annually to the Columbia River system. By the early 1970s, the return of wild fish was less than ten percent of that, according to some estimates.

While interruption of migration paths and the commercial harvest of salmon took a toll, a huge impact on the Middle Fork wild fish population may have been a result of irresponsible mining. A former editor of the Idaho Statesmen wrote:

> *"When Porter Brothers Corporation began dredging the uranium-rich black sands of Bear Valley Creek in 1955, it vowed to keep the pure high-country water 'crystal-clear', to protect some of the richest salmon and steelhead spawning beds in the entire Snake-Columbia system...*
>
> *..the devastation was incredible. Miners diverted creeks and tributaries into canals and dredged channels out of existence. They shipped the radioactive ore to a mill in Lowman, Idaho, for processing. Spawning gravel was scooped out and dumped. Forest cover was stripped.... radioactive sediment was sent streaming down the creek, suffocating spawning beds, clogging rearing pools, and muddying the Middle Fork downstream.*
>
> *Three decades after mining ceased, the ponds, tailings, and sediment remained while most of the fish did not."*

As a result of widespread abuse of public lands and trusts, and the destruction of wildlife and natural resources, the 1950s and 1960s witnessed a raising of conscious awareness within the country regarding our habits and their impact on the environment. The Audubon Society, Sierra Club, and National Wildlife Federation were becoming more radicalized after government failures in food and drug, agriculture (pesticides), and natural resource management reached epic proportions. The pendulum began to swing, and environmentalists pressed the government to implement meaningful changes. Unfortunately, the government was no more effective at solving problems than it was at avoiding them. But the bureaucracy moved in response to public pressure and, if nothing else, operated with more sensitivity to the national mood.

The Wilderness Act of 1964 provided an acceptable legal definition of wilderness in the United States. It was legislated to establish a National Wilderness Preservation System for the permanent good of the whole people, and it initially protected 9.1 million acres of federal land. The result of a determined effort to create formal criteria for designating federal wilderness and mechanisms to protect it, the Wilderness Act was signed into law by President Johnson after over sixty legislative drafts and almost nine years of work.

The Wilderness Act was well known for its succinct and poetic definition of wilderness:

> *"A wilderness, in contrast with those areas where man and his own works dominate the landscape, is hereby recognized as an area where the earth and its community of life are untrammeled by man, where man himself is a visitor who does not remain."*
> *- Howard Zahniser*

The Wilderness Act not only served the public by providing a framework of legislation to protect its most sacred natural resources, but also to remind the public of its own responsibilities to protect and preserve the wilderness. Even the vacationing public was morally obligated by The Act to leave the wilderness the way they found it.

Additionally, the '60s witnessed the explosion of environmentally-conscious movements and organizations: the World Wildlife Fund and Friends of the Earth were established. The Trails Systems Act, Wild and Scenic Rivers Act (which included the Middle Fork), and Wildlife Refuge System Act were legislative reflections of the national mindset. The Environmental Defense Fund ('67) and the National Environmental Policy Act ('69) were the obstacles presented to those who would resist that mindset.

One solution to migration obstacles posed by the dams was the installation of fish ladders, or fishways. Fish ladders represented compromise; they were commonly used to provide native fish with an alternate path upriver past obstructions, and the dams could stay in place. Designs varied depending on the obstruction, river flow, and species of fish affected, but the general principle was for a ladder to consist of a series of ascending pools reached by swimming against a stream of water. Fish leapt through pools, rested, and repeated until they were able to escape the ladder. The bottom-line benefit of fish ladders was questionable, but they were better than nothing. The Forest Service determined that the salmon struggling at the end of the trip up the Middle Fork needed help in their efforts to reach their traditional spawning grounds. Somebody proposed that a fish ladder be constructed at Dagger Falls, and a budget was approved for the project.

Unfortunately, there was no way to transport the materials required for the project without a road into the backcountry. As a result, a road was built from Bear Valley to Dagger Falls by way of Boundary Creek to allow for the movement of the materials needed to build the fish ladder. The project was completed in 1960, and the Middle Fork River was never the same. An unintended benefit of the Boundary Creek road was improved access to the river for commercial outfitters and the general public. It allowed boaters to skip the punishing rocks of Bear Valley Creek and, essentially, the first ten miles of the river, and put in where the water 'got good'. The improved access led to a dramatic increase in recreational usage.

Before the road was built, it was a hassle to launch a trip down the Middle Fork. The early boaters launched at Bear Valley Creek, a few miles up from where Marsh Creek joined to form the headwaters of the Middle Fork, and they encountered rough running from the outset. The bottom of the river was close and rocky. One of the first groups of men to successfully run the river - the Colorado River Club - remarked that "the creek was one continuous rapid, and too shallow and rocky to get boats through without pushing and leading." The CRC members aborted their first attempt after spending four rough days on the river to reach Dagger Falls and finding it impassable. They hired packers to portage their gear and boats around the falls, and returned the next year to complete the run. Years later, Prince Helfrich and Woody Hindman would camp with their guests two miles above the falls, and row down early the next morning to use a block and tackle to lower their boats to where they could be carted to the base of Dagger Falls. There was no easy way to get into the good part of the river. In that first decade, only a few dozen made the trip each season. Growth was incremental; only a handful of outfitters offered the trip, and their marketing was mostly word-of-mouth.

The Forest Service road to Dagger Falls opened the area up and triggered a dramatic increase in recreational use. Between 1962 and 1970, the number of Middle Fork floaters increased by a factor of five.

1962 :	625	Middle Fork floaters
1964 :	753	"
1966 :	1260	"
1968 :	1529	"
1970 :	3028	"

(Source: U.S. Forest Service)

By 1970, there were more than 3000 boaters running the river each summer, and the number had doubled from two years prior. The large increase in recreational use on the Middle Fork created considerable concern on the part of the Forest Service regarding the impact of the increase on the environment. In 1968, they were granted authority to take ownership of the problems when the river was designated as 'wild and scenic', which put the Middle Fork under the Forest Service's control. Even at that time, there was some friction between outfitters and the general public. There were problems caused by all the people on the river; mostly trash disposal, toileting, and general neglect. There was a perceived lack of professionalism and general safety (which became a bigger problem in 1970, when three people died on the river; one at Sulfur Creek and two at Weber Rapid, a trip in which the newsman Tom Brokaw participated).

In an effort to formalize a perspective on the problems, a study was initiated in 1970 to determine the 'recreational carrying capacity' of the Middle Fork. It was part of a Wild and Scenic Rivers project funded through the Water Resources Research Institute at the University of Idaho. The objective of the study was to define problems associated with increased usage of protected resources and to suggest solutions. It concentrated on three areas of interest:

- Legal constraints described in the Wilderness Act (1964) and Wild and Scenic River Act (1968)
- Physical constraints dictated by the resource and its inhabitants
- User Desires (defined by users surveys with questionnaires)

Legally speaking, the Wilderness Act and the Wild and Scenic River Act provided only general guidelines for usage and preservation. It left too much to the discretion of agency administrators, which resulted in problems of interpretation.

The physical restrictions or limitations of the area, those characteristics upon which all other resources intimately depended and which ultimately must be protected, were the ground and water and life within its boundaries. The Middle Fork drainage was safe habitat for numerous species of fish and wildlife. It was home to black bear, elk, deer, goats, sheep, wolves, mountain lions, and eagles. The river was an important spawning ground for salmon and steelhead. The protection of the habitat was the underlying priority in every question and consideration. The public's use of the protected resource must not result in undue impact on wildlife, flora, or fauna.

'User Desire' was the most subjective section of the study. It was encouraging that the Forest Service asked the public's opinion. They took a sample of Middle Fork boaters and personally interviewed them in the summer of 1971. Additionally, a mail questionnaire was sent to registrants after that summer season. The questions in the survey addressed demographics of users, their attitudes toward a managed resource, and an 'experience assessment' regarding their trips down the Middle Fork River.

The survey taught the Forest Service that people ran the river for solitude, scenic attractions, its primitive atmosphere, and whitewater adventure. Their response to survey questions was surprisingly consistent with the criteria outlined in the Wild and Scenic Rivers Act and Wilderness Act as being important to the recreational use of both. It was the combination of all those features that made a float trip down the Middle Fork a memorable and unique experience.

In the course of conducting the study, it became obvious that a plan for managing the river was necessary. While interviewing river users, it was determined that one of the ugliest impacts of increased human presence on the Middle Fork was the accumulation of garbage and debris. Some users behaved so irresponsibly that the

Forest Service was forced to create a river patrol to pick up garbage. The solution became less effective, however, as the pounds of garbage increased faster than the number of people. The collection of litter represented a substantial expense and, in 1971, a carry-out regulation was instigated. It was a problem so big that it might have constrained further use of the Middle Fork.

Middle Fork boaters were asked to specify positive and negative highlights of their trips. Boaters were experiencing a variety of problems associated with general congestion. The use of the river by some was adversely affecting the wilderness experience of others. Sources of potential conflict included large parties, campgrounds, and aesthetics.

Party size was self-governing with small groups. They were private affairs into the wilderness, and lost some of that wilderness mystique when there were too many people in camp. However, according to the survey, approximately 70 percent of the sampled boaters employed the services of a commercial outfitter to run the river. Occasionally, those groups grew to considerable size; some rumored to have exceeded one hundred people. Users felt group size could be an important problem as interest in the river and wilderness increased. Large groups of people, even for a relatively short period of time, could severely damage a backcountry site, and simultaneously serve as a disappointment to those seeking a secluded wilderness experience. Two-thirds of respondents felt that party size should be managed.

It was a difficult problem to address, and the constraints were great. More than 80 percent of the boaters surveyed floated the river in July. The thousands of people that visited each year had to be accommodated during a very tight window in time; one governed by Mother Nature, and completely out of the Forest Services' control. For all practical purposes, the boating season on the Middle Fork was from the time the road to Daggar Falls opened (late June) to when the flow at Daggar Falls dropped too low to be safe (early September). Eight to ten weeks to move thousands of people through the middle of nowhere while maintaining the wilderness experience. Group distribution would have to be managed.

One of the least desirable experiences for boaters occurred if they were forced to camp with other parties. Most people did not want to camp with people from other groups. The availability of campgrounds, especially on the lower portion of the river, could be a limiting constraint to boater use in the future. The lower portion of the river was relatively narrow and the number of available or suitable campsites was limited, especially when high water covered sandbars that were used during the latter part of the season. There could only be as many people on the river as there were campgrounds for their bivouacs, and that was an unchangeable physical constraint. Campground usage would have to be managed.

Almost without exception, Middle Fork boaters indicated that one of the major reasons for floating the wild and scenic river was to observe the scenic attractions of the area. The legislative guidelines - that man's imprint should be minor - as well as the desires of most boaters dictated that usage levels should be sufficiently low that recreational use did not degrade the system.

The case study of the Middle Fork was the first step toward a Recreation Wilderness Plan that outlined constraints that might ultimately limit use of the Middle Fork and, more importantly, a plan for managing usage within those constraints. There followed a reorganization of the forest boundaries; the Middle Fork Ranger became responsible for boating operations on the Middle Fork, and the North Fork District of the Salmon National Forest administered the main Salmon River. In 1973, the "Recreation Management Plan, Middle Fork Salmon Wild and Scenic River" was completed.

The important details of the plan as they related to recreationists were consistent with survey respondent concerns:

- ✓ Starts limited to 3/daily for outfitters, 4/daily for private groups. Group sizes were limited accordingly. Trips were by permit only; outfitters were 'grandfathered' into the permit allocation system with Special Use Permits, while private groups applied for dates determined by a computer-generated lottery. The objective of limiting starts was to distribute usage evenly throughout the season.
- ✓ Campsites were distributed before daily launches, and each group received assignments consistent with their size. Requests were accommodated, if possible, but not guaranteed.
- ✓ Strict regulations were put in place regarding the removal of all human waste.
- ✓ Strict regulations were put in place regarding the use of fire pans (and prohibiting the construction and/or use of fire pits).
- ✓ Safety 'regulations' were suggested, outfitter liability insurance was required, and licensing requirements were established for professional guides.
- ✓ Launch site (Boundary Creek) and take-out site (Cache Bar) were constructed to facilitate easier river entry and exit.
- ✓ Special Use Permits were issued to 32 commercial boaters who had valid outfitter licenses issued by the Idaho Outfitters and Guides Board. The special use permits provided for temporary or transient land occupancy and were issued only on an annual basis. The Challis National Forest issued permits to the following commercial float boaters during their first year of administering the Middle Fork of the Salmon Wild and Scenic River:

Don Hatch	David D. Giles
Dave Helfrich	Leroy Pruitt
Dick Helfrich	Bob and Jay Sevey
Val Johnson	Don and Bob Smith
Kenny King	Roy Taylor
Vladimir Kovalik	Eldon F. Beam
Mackay Bar Lodge	Dave Burk
Kenneth E. Masoner	Jim Campbell
Hank Miller	Martin Capps
Stan Miller	Bob Cole
Dale Ney	Clifford O. Cummings
Roy Nicholson	Jack Currey
Mel Morrick	John Dorr
Louis A. Elliot	Omer Drury
Henry Falahy	Guth Enterprises, Inc.
Jay D. Foster	Eldon Handy

The Forest Service modified the districts to make managing the new area more effective. With the new District boundaries and workloads, it was decided that the duties of the Middle Fork District and the Middle Fork River Rangers could be combined. Since Challis was geographically centered between the two ends of the Middle Fork and right next to the Salmon River, it was selected as the District Office. This was accomplished in July of 1973. To manage it all, Ted Anderson was appointed River Manager and reported to Jack Bills, Supervisor, U.S. Forest Service. It was the beginning of twenty years of service for Andy Anderson's son.

"You know, I can't remember the exact date that Dad sold the outfit to Eldon Handy," Ted reminisced over another lunch at the Y-Inn. The special was a Chicken Parmesan sandwich. There were green-and-white checked tablecloths on the tables. Matt from Fish & Game sat down at the table next to us with the sheriff. I nodded 'hello' to him, but was listening to Ted.

"It was during the time when they decided to start the permit system, and there were a limited number of outfitter permits. What was happening was Dad wasn't running the river anymore. He was living in Challis and spent more time working his claim at Yankee Fork. Some guys would get into the business by buying up an existing outfitter. If you eliminated that outfitter, you could put in for his permit. Handy had a partner, name was Cavanaugh… something like that. They had a business, and Handy bought out my dad. That was the only way to do it. There were only, I don't know, thirty-two outfitters? But before that, you didn't have to have a permit. You'd just get on the river.

And that's what happened, you know; that's the real story. Some Forest Service guys from Utah went over to Dagger Falls on 4th of July one year, and there was a line of cars almost 3 miles long waiting to get on the river! That's when they actually decided it needed to be managed. And that's when they began the process of putting together a river plan with a permit system and how many outfitters they could handle."

"Back in the early '70s."

"Yeah, that's right."

"Did that cause a lot of trouble?"

"It caused a lot of trouble between the professionals and the privates, yeah. The outfitters used an 8-day launch cycle, so every eight days, they'd start out again on the river. You had all these outfitters and they were all lined up… and that was after you started managing them. Before that, you had a real mess. No campsites assigned, so especially in the Canyon, you'd end up with 60-80 people on one of those little old sand bars. The season used to start on June 25th, but then extended it 'til it was earlier in the spring. Higher water. Even less room.

Then they increased the party size. Somebody else decided on that; not anybody here at the District. And that was a bad deal, when they increased the size of the groups. Now you had a lot of places overcrowded, especially Boundary Creek. Number of boats and all that, trying to put into one spot.

Then they went to the pack out system. They had one for trash when I went to work for the Forest Service; I think they started that in '64. Then we got into the fire pans. You'd go to a campsite and everybody wanted their own fire ring. So there was just one fire ring after another… and they were always dirty because people would throw their trash in there, too. So we got into the fire pan thing.

And then after that, we tried the toilet thing. That's when my brother Terry went down and took those two fellas. One of them was Finance and the other was Recreation. That's when they built the first one. We hauled that down. They put in garbage pits. They put a wooden thing over the top of them, but the bears would get in. They got rid of that idea; everybody could just haul their own trash out. I tell you something; the original crews hauled a lot of trash out of there. Left to their own devices, people would ruin that place."

"So you stayed here in Challis? You lived and worked right here?"

"Yeah, lived in the same house. Worked down at the end of Main."

"How many people were in the Forest Service facility back in those days?"

"Rangers were living there. But the Supervisor's Office was the bigger building. Until they built the new one, and then they built that new rangers dwelling right next to the Supervisor's Office. There must have been a permanent fifty or sixty people."

"But now…. is anybody living there?"

"Mm-hmm. They use it for their summer help."

"But no one lives there full-time. Not anymore. But they used to…"

"Yeah. The Ranger used to, and the Assistant Ranger."

"What happened? I mean, the river is as active as ever. Where's all the staff?"

"Well, it happens because they reorganize. They join stations sometimes, like when Yankee Fork joined up with Challis Ranger District… consequently, the one up the river by Clayton became a workstation. And sometimes they just cut budgets. It's all in the name of saving money, until after its done and they say they didn't really save any money.

When I wasn't on the river, that there was my office. I was there 8-to-5 and everyone knew where I was. And I was in that office during the winter, when we went through the permits process. When people used to call me and cuss me, and tell me I was playing favorites."

I could tell it bothered Ted, even after being retired for so long. He was an honest man; a man of integrity. His loyalty was to the river, and his job was the fair administration of the rules laid out in the Recreational Wilderness Plan. He worked long hours, management hours, often giving up a normal life for the privilege of working on the river.

"Even on weekends, I'd be working to get supplies ready to take on some other trip. Other than those trips with big shots, I took water samples along the river several times each season, and I'd do inspection trips. And I'd come home and my family would all be waiting for me to take them camping."

I knew for a fact that he took them camping. There were pictures on the walls.

"I had one ranger that I didn't get along with. She was my Ranger. Middle Fork Ranger – first female ranger in the area. She tried to get me fired all the time. My friend Burl Hamilton got along with her better; she liked horses and 'Bud' was a packer and worked the trails in the district since '68, so they got along just fine. But she didn't know anything about the river, and she didn't want to know. I knew she was just passing through, so I was able to deal with it. Just say "Yes, ma'am" and "No, ma'am", and then she's gone. But I stayed. I was the River Manager for twenty years."

The Recreation Management Plan, Middle Fork Salmon Wild and Scenic River was implemented in 1974, the same year that Ted Anderson was employed by the Forest Service as River Manager for the new district. The timing was fortuitous; 1976 was a record year of recreation on the river. Almost ideal water conditions from June through August resulted in over 5,800 users, which represented a 24% increase over the previous year. By the time Ted retired from his position twenty years later, that number doubled and surpassed the hypothetical 'river carrying capacity' determined many years before. He was uniquely qualified for the job and, most importantly, ended up where he belonged. The wilderness had been part of his life since the day he was born. He worked in the backcountry as a packer and guide until he was in his early twenties, then spent twenty years working around town. Just when it needed him, Ted dedicated the next twenty years to serving as the Forest Service River Manager for the wild and scenic Middle Fork.

Big Shots

August 23, 1978 : (informal exchange with reporters following President's Raft Trip – from the *Weekly Compilation of Presidential Documents ; President Jimmy Carter*)

"I'm glad to be in Idaho. I used to come out here back in the early fifties when I was working under Admiral Rickover, came into Pocatello and Arco. And I came out here when Cecil Andrus was Governor – we're very close friends – and I'm really looking forward to going down the Middle Fork. I'm an old kayaker and canoeist."

"You better watch out for Andrus, he's a fly-fishing enthusiast. Is there a challenge to do any particular contest as to how many and how big you're going to catch on the...."

"He brought his own fly-casting outfit. I left mine back in Georgia, so I'm going to have to borrow one."

[laughter]

"Do you see this as an unusual vacation for the President?"

"Well, I understand that it was an unprecedented vacation, but I've been looking forward to this for a long time. Cecil Andrus *and I have been close friends since we were elected Governors, and he's never lost an opportunity to tell me about the beauties of Idaho. And my own background in whitewater canoeing and kayaking has prepared me to want to come out here and enjoy the Salmon River. So I was looking forward to it. And even if it was unusual, I think I benefited from the fact that I decided to do it."*

"What about the youngest member of the family staying back?"

"Who, Amy? Amy was here with us."

"How was the trip, Mr. President?"

"The best 3 days I ever had. It was really great."

"What were the high points?"

"We saw mountain sheep on part of the trip. We saw golden eagles, and we caught a lot of fish. We had a contest today. I got 59 trout."

"Did you get hung up on any rapids at all during the two days?"

"No. We had a little trouble on one waterfall."

"That was Tappan Falls. We were there."

"I thought maybe they changed the name by now."

[laughter]

"Did you go in the water at all?"

"By accident, no."

"Have you missed what has been going on in the world?"

"I talked to the VP on the phone this afternoon, got a report on several things. I have gotten a Presidential Briefing every morning at 7am from the State Department and also from the CIA. So we had good radio communications from the outside world. Everything seems to be quiet".

"Did the Interior Secretary lobby you at all on the wilderness issue up the river?"

"Well, he gave me a lot of information about it. But we will have to discuss it very carefully. I think one of the best things about this region is that right through the part we went, it ought to be preserved and not destroyed. But back off from the river where we saw flying in from the plane, were very good productive timber regions where we ought to harvest timber and let the country benefit from it. So, I like a good balance between preserving the natural beauty, unchanged, on the one hand, and harvesting growing timber in appropriate areas, which is what he is working on, and I agree with him."

Trailed by members of the press, President Jimmy Carter, his wife Rosalyn, sons Chip and Jack, and daughter Amy embarked on a float trip down the Middle Fork of the Salmon River late in August of 1978. They camped each night on the river banks, fished, enjoyed a fine display of wildlife, swam during the day, and marveled at the pristine beauty of the remote area. The trip progressed smoothly with the exception of a sweep breaking on the raft in which the President was riding. The party landed on the river bank and docked for an hour while Norm Guth, the outfitter and guide, made the necessary repair. The rest of the trip continued without incident.

Department of Interior Secretary Cecil Andrus, responsible for promoting the trip, accompanied the Presidential party. The purpose of the trip was for President Carter to get a first-hand look at the Salmon River and the Wilderness Area, and better understand what could be opened for development and what needed to be protected.

It was not like any other raft trip organized that summer. A great deal of preparation went into every Presidential visit. Security was paramount, and it must have difficult to prepare for a Presidential visit to the backcountry.

Ted explained…

"When the big shots went, like Peanuts, they went with a private outfitter. But the Forest Service organized all the pieces, and I represented the Forest Service. Made sure everything was put together, predictable. Radio, Secret Service, White House people, things like that. I didn't actually serve as Guide to the President."

"So you did all the prep work. Did you have much time?"

"About a month. It was about four trips on the river before the big shots ever went down. White House people wanted to look at it first. Secret Service had to look at it. I was going around and around and around. My job was all about their prep. The actual trip with Peanuts and his family was only a few days. With Bush, it was just a day trip. Just a Vice President; not as big a deal."

There was something characteristic of Central Idaho in his description of the trips. Ted was honored to have participated, but it did not define him or his life. He was no more impressed by dignitaries than his friends were impressed by him when he walked into the restaurant for lunch. A lot of people in Challis knew Ted, but not as part of the special river family; only as one of the guys who grew up around there. He was proud of his expertise and experience, but not overly impressed with himself. He was typical of Challis and the people I met there. They were easy people to like.

"You should have seen it when the Secret Service saw those two guys hiking down the river, when we took those guys with Carter down. That caused a commotion, I tell ya'."

"What's that?"

"We were there at Elk Bar and these two guys had hiked down one side, got a ride across the river, and come hiking back on our side. Boy, you talk about Secret Service getting on top of them."

"There's no trail on that side."

"I know! They were just climbing around… boy, they really picked a bad day to hike around the backcountry. Those Secret Service fellas really checked 'em out."

"I remember when they came out at the end of the trip. It was so cool!" Tammy joined us for lunch; she spent as much time as she could with her Dad. And the special that day was Mexican food – her favorite.

"We all went down to the mouth of the river to see them," she added. "And we got to go aboard Marine One! There in that field by Owl Creek. It was very exciting. I was like maybe a sophomore? A junior?"

She was so proud of her father and grandfather. Tammy collected pamphlets and magazine articles about the Bar-X Vacation Ranch from years before she was born. She loved stories of her Grandpa Andy and his exploits in the early days; she was his biggest fan. Other families' names reflected generations of ranching history in the area; big names well-respected in the community. Her family name had come to represent the river and the wilderness that surrounded it. They weren't known by everyone; only by those who knew the river, and Tammy rightly honored her family's special contribution. She was the archivist; the keeper of all the best pictures from the days when Andy and Joe worked the whitewater. She was proud of her family's legacy, yet she chose a different path. She did not work in the backcountry, but went to the mines instead.

"So what is your story, Tammy? You're the black sheep of the family. You chose the mines over the river?"

"Well, my grandpa was into mining," she corrected me with a smile. "He did that up until the day he died. Grandpa Andy was always into mining, so I stuck with that part of the family tradition. He had claims up in Yankee Fork, and I spent a whole lot of my childhood up in that area. And that's how I met Corey, my husband. When Thompson Creek had their first layoff, I went to California to work in a gold mine and that's where I met him. But I started at the Man Camp, right here in Challis."

"The Man Camp?" I asked. It did not sound like a place a father would want his daughter to work. I looked over at Ted, who nonchalantly dipped his chicken quesadilla in sour cream and listened to his little girl share her side of their story.

"It was a place for all the contractors that were there putting in the mine," she replied. "I was a waitress. It was good pay, and a friend of mine and I did that for a summer. It was crazy. Then I went right from high school to go to work at the mine."

Tammy had happy blonde hair, like her mother's. She looked like her mother. Ted was the only dark-haired one in their house. Tammy and her brother Mike got their noses from Dad, but their eyes and smiles were straight off Mom's cheerful face.

"I went to work at Thompson Creek in 1983. It was real good money. They were mining molybdenum. A 'molly' mine. I started out driving a truck, as a casual. Then the mill was getting ready to fire up and my brother Mike was up there, and I got on permanently as a tailings operator. I worked in the mill and packaged, did pretty much everything back in the mill except run the control room. It has its own education system – on the job training. And once you got experience, you could take it with you. That was the great thing. I spent the last twenty years working in labs, even though I'm not a metallurgist or chemist – we have those – but I'm experienced and therefore educated."

Tammy was always working. Even when the mine closed down, she found work around town, or in another town. She could seemingly do anything, because she always had a job.

"When they had their layoff in '86, I moved to work in a gold mine in California. The guy who was my supervisor and his girlfriend, they were going and I went, too. Lived there for a while, and that's where I met my husband Corey. I brought him back to Idaho and, when we got to Stanley, he was like 'Wow! Why did you ever leave this place?' And we've been back here ever since."

"And you raised your family here. Your daughter lives here. In Idaho, right?"

"Yes. Which kinda completes the circle, really. Shauntee worked at Boundary Creek, managing the launches. Now she wants to stay here and work in the forest and on the river. Back where it all began for us."

"And you just let Tammy do what she wanted" I asked her Dad, who sat across from me. "Why didn't you steer her to the river?"

"Well, it was her life and I let her live it," Ted winked where I could see it, but kept a straight face to his daughter. "And when she left, she says 'I am never coming back. I hate this town. There's nothing to do here!' Of course, a lot of small town kids say that. And you would be surprised how many of those kids have come right back here to home."

Ted never left Challis. He went away for a year or so when he served his country between the wars, and there was a period of time in his early years when he did not work in the backcountry. But he never left Challis, where he moved as a child with his family to pursue his father's dream. He packed rafts through the largest wilderness in the Lower 48 when he was just a teenage kid, and packed a President down the river thirty years later. He lived through the Great Depression and two World Wars, but he was just another guy at the Y-Inn in Challis. And he liked it like that; he was admirably content with his life. I respected and admired him for his accomplishments, and I learned to love him for his humility.

Nowadays

I strolled down Main Street to Ted's house, and the air was alive with summer. Cottonwood seeds were carried in the wind, so light and floaty that a slight breeze blew them sideways through the air. They looked like bees swarming through the trees. The cotton only dropped for a couple of weeks each year, but when it dropped it was thick in the air and looked like pale roe where it piled up at the base of trees and rocks. Along with the lilac blooms filling the air with their perfect fragrance, the flying cotton flagged the season.

I rented a house in Challis that summer to finish the book. It was a little two-room cabin on Challis Creek, two blocks north of Main and three blocks west of Ted's house. It was perfect for the project; located on a lot across the blacktop from a corral, with a creek down one edge of the property a hundred feet from my porch. I kept a lawn chair in the trees at the water's edge and worked outside when the weather was good. There was no traffic; just a vehicle or two at a time and the same people every day. Everything I needed was close at hand, and most of it was open on Sunday. It did not feel like a thousand people lived there; it was too peaceful to be that big.

I was going to Ted's house to meet his granddaughter, Shauntee Rice. She carried her father's family name, but was 4th generation

'Idaho Anderson' and, like her mother, proud of her family's past. She worked and lived in Pocatello, and was preparing for a career in the Forest Service.

We were gathered in Ted's living room. He was sitting in the corner next to the front window. Shauntee and I were across from him on opposite ends of the couch. Ted had become comfortable with the whole process by that time. He was smiling and ready to engage, but sat quietly as Shauntee told her story.

"How did you get the urge to work on the river?" I started. "How did it skip your uncle and your mom? And why didn't you follow your mom into mining?"

"I had the opportunity to apply for YCC," she answered, "the Youth Conservation Corps, which is where they take you out and show you what the forest is all about. You make slash piles, you live in the wild, and you learn on the job what the Forest Service is all about. They had taken me up to Boundary Creek one time, and obviously I knew about its history... and at the end of the season they asked me where I wanted to be stationed if I stayed on. I said, 'Put me up at Boundary Creek.' And that is exactly what they did. But the first thing they said to me was, 'are you sure you want to do that?'"

"That's the worst job on the river," Ted exclaimed. "Trying to balance things between the privates and the commercials. Everybody wants more than their share."

Ted was openly proud of his granddaughter. Part of it was because she carried on with the family's service to the river, but part was pure appreciation for her courage. She was articulate, confident, and determined. I'm sure she used the same "fair but firm" approach that made Ted so successful in the job.

"Did they know who you were when you asked for Boundary Creek?

Did they know that 'Shauntee Rice' was Ted Anderson's granddaughter?"

"Oh, yeah. Sure. There were pictures of Grandpa Ted at the Forest Service, on the walls, all over the place. There are albums of pictures. Everybody knew him, even twenty years after he retired! And it wasn't a job that people wanted. It was a hard job. And I had just graduated high school – I was 18 years old. My first 8-day hitch, my parents were saying that I would come back loving it or hating it… and I came back loving it. I worked it for six seasons."

Shauntee did not look old enough to have worked the river for six seasons. She had that bright vibe of young people completely caught up in college, but in fact she was already out in the working world.

"Were there any grouches? Did they give you a break because you were young, or did they give you a hard time?"

"Oh, lots of grouches!" she laughed. "Especially with regards to campsites; that can get a little heated. There is this one old guy that Grandpa called 'Peg Leg'."

Ted snickered at the good memory.

"A nice enough guy. Grandpa called him 'Peg Leg' because he had an artificial leg," she continued, "and he used to give me a hard time. A real hard time. But I guess one day Grandpa ran into him in town and asked ol' Peg Leg why he was giving his granddaughter a hard time. Things changed after that, real quick. A total 360. All of the sudden, it was like 'Oh! Your Ted Anderson's granddaughter!' Now ol' Pegleg is my buddy."

"You worked six seasons on the river, and then left?"

"I started in 2012 right after high school and worked there for six seasons. By that time, I completed my degree in Wilderness

Management. I went back and applied for a seasonal job this year, hoping to work back into managing the resource, and carrying on the tradition."

"What is a degree in Wilderness Management?"

"It teaches you how to manage people in a wilderness."

"People? You're kidding. That's not how it sounds…"

"Yes. How to manage recreation, understanding the impact that people have on the resource. I took classes at the University of Montana."

"Was it redundant? I mean, you kinda grew up with that."

"A little. Truthfully, my background put me in a position to assist with instruction. But there was a lot of new material. The legal aspect of things was a big part of it. NEPA. How to keep the public involved, but without a permanent presence. The education helps, but my real value right now is the experience I have dealing with people.

Boundary Creek is not a good place for bad news. People get mad. They had to draw for a permit. They are issued a permit for a specific date. It costs a lot of money. And now they get bad news? That is hard to manage, and they can't teach that in school."

"What kind of bad news?"

"Oh, like blow outs. One year there was a bad blowout just above Sheepeater and there was a blockage in the river. The Manager came up to sit with me and tell people what was going on, so they could decide to wait or cancel. Fires cause problems, too. Even then, all you can do is advise. A lot of people will run it anyway, and pull out above the problem, portage around it, and continue. People are that committed."

"Even in high water, you can only advise people," Ted agreed. "You can't tell them not to go. Because the minute you say, 'Ok, now it's safe to go', someone flips over and says, 'You told us it was okay.' And you got a lawsuit on your hands."

"Can you make a career out of this?" I wondered. "I mean, you're a modern girl. Can you make a living working as a ranger? Does the Forest Service... the government see it as a career? Or will it always be a summer intern kind of thing?"

"I think they see it as a career. Certainly they do. There is a new ranger managing this year; she's in it for the career. She has been asking me questions about the job and the river. She wants to know, and we work together. My background gives me a leg up; I believe I can work my way up to District Ranger. I believe there's still a meaningful career in the Forest Service."

"When was your last trip? Have you been on the river lately?"

"Every year, you get one trip down the river as a Forest Service employee. You check out the campsites, all that stuff. It's really cool. It gives you the experience you need to talk to the people at Boundary Creek about the conditions that year. Those kinds of things."

"How do people behave nowadays? We are more educated, right? They don't trash the river and the wilderness area anymore, do they?"

"Well... hmmm.. yeah. They do. Not all or even most, but yes. Sadly. There are times when people actually leave trash in the campsite. Garbage in the firepit... unburned trash. Very unsightly to the next group. Not to mention the occasional fire that they cause.

Outfitters and some privates end up carting out trash left by others."

"Which is sad," chimed in Ted. "Because if you trash it out, what have you got? What has anybody got anymore? It is an education problem. And some people just have that personality; they're there to have fun and someone else can clean up the mess."

"How does your normal work week go out at Boundary Creek? Do you drive out every morning?"

"Oh, no. The gig is 8 days on, 6 days off. Basically, its Wednesday to Wednesday. They have a manufactured home at Boundary Creek. Its 3-bedroom 2-bath. Super nice. Fully furnished. Two 'fridges, so lots of room for food."

"Do you bring your own food?"

"Yeah. They don't pay per diem because it is our work station. We're not out-of-town. If they fly us out to Indian Creek, then we get per diem pay…. only while we're on the river. And we get paid during our days off. So it's all good."

"You get paid for days off?"

"Not exactly, but it feels that way. You get paid for an 80-hour week. It nets out like any other full-time job. And that's what matters. It's a living; you won't get rich, but you can make a career out of it."

"We used to use trailers," Ted remembered. "Old trailers, but decent enough. Then we got an engineer down there, sorted out the water. Got a good water system in… and then went looking for a good site. The problem is that 'good sites' in the backcountry have been good sites forever. And the site we picked turned out to be a popular site for Native Americans back when, so we couldn't use that. But we got it figured out and now have a nice place down by the river."

They smiled at each other. They both worked at that place.

"Do you remember Andy, Shauntee?"

"No. I was born a year before he died."

"But you remember Grandpa Ted when he worked on the river…"

"Oh yeah," she grinned. And he grinned back.

"Was it a special thing, or just his job?"

"Oh, definitely a special thing. Even after the fact, you know, when I saw all the stuff in the hall, and all the stuff out in the garage."

"Well, where I'm going with this is…. Your mom didn't take up the job. Her brother didn't either. But you did. Was it because of what you saw in Grandpa's work?"

"It was opportunity, really. I knew about the river and knew it was special from growing up around my Grandpa. And our family's history. But if the opportunity to sign up with YCC hadn't come along, I don't know that I would be working there."

"I guess it worked that way for Ted. He left the river for a long time, worked at mines and other jobs. It was years before he returned to the river, but when the opportunity came up, he did. And he stayed there. He retired there."

"With my family history, our experience, I feel like I have a role to play here. When new rangers come into the district now, they don't come with the same backgrounds as they did thirty years ago. They really don't know that much, and those of us with experience serve as educators – not just to the public, but to our peers, as well."

There was a maturity in Shauntee that came out when she talked about the work. I was still talking with the same young woman, but there was a subtle change in her demeanor, a certainty that her intuitive knowledge would be there when people needed it. As if she inherited an Anderson Masters in River Management.

"When I was working, I was just a Technician," interjected Ted. "I didn't have a degree. But I stayed on the Middle Fork for 20 years, and I learned. Now, these young people move from place to place. It's hard to acquire the expertise you need during a relatively short assignment. But with the experience in place, you had this nucleus there all the time. So when a new ranger comes in, Shauntee can supplement his education with knowledge about the area, about the specific practice or resource. All the old guys with all the experience are gone now, and that is not a little thing."

He was right. I had seen a changing of the guard in the few years I'd lived in the backcountry.

"Old Bud Hamilton used to work the fires and the trails, he was in Salmon – he's retired now. I'm retired. Jim Upchurch retired last year and took all his expertise with him. Thirty years in the backcountry; nothing he hadn't seen. These are important resources, and continuity is needed. That's the value of someone like Shauntee. She can sit with them and think with them… and say 'We've already tried that' or 'That's a new thought'. Saves them time."

"And when some of those people leave, the Forest Service crews they trained leave, too," I added. "Like Upchurch; when he left, three very experienced and capable individuals left, as well. They didn't want to mess with the changeovers."

"There are those that come in and do not want the 'little people' to tell them what to do," Shauntee explained. "Big ego trip. They're not all like that; this new lady that interviewed me…. she said that she would probably end up asking me more questions than I asked her. But you definitely get those people who come in with a plan to change everything."

"What is the profile of these people? Is there any particular pattern to the know-it-alls? Are they college educated?"

"Oh yeah," Ted chuckled. "That's the problem. Bud used to call them 'overeducated enthusiasts'. They come from cities and have an interest in Nature, go to school for a few years, and think they know everything when they first set foot in the wilderness. And of course, no disrespect intended, they know next to nothing. And their college-cultivated egos have a little trouble with that."

"The last one for me was a lawyer," Shauntee related. "He was a lawyer before he decided to become a forest ranger."

"And that might tell you something," I laughed. "How successful was he as a lawyer if he quit to go to the woods."

"I'm not saying anything is wrong with that," replied Shauntee, "might be everything right about that. But his ego filled the room, and he did not know what he was talking about. He was hired because of his education, but he had no education in forestry or wilderness management. It should be a career. We need career people managing our wilderness resources, not hobbyists. Grandpa taught me early on that I would only be as good as the people working with me. He told me to listen to them; let them get involved."

Ted really opened up with Shauntee around. It had become less of an interview and more of a conversation. He shared with enthusiasm.

"I was lucky because Bud & I got along with most of the rangers we worked with. Back then, they saw us as the anchors in the team.

Except for Mrs. Fox. Bud & her got along pretty good; she liked horses and he was a packer. He could get keep her busy with stuff that made her happy. She wasn't so much into river stuff. That was the problem with that last guy; he didn't know anything about the river and didn't want to know anything. But it was part of his job."

"The main thing about this job," Shauntee concluded, "is teaching people that we have to stay within the lines. You can't let this group take some extra people because you like them…. That just screws up everything, including your credibility. You have to be firm."

"But that's okay with you. You don't mind the job. Clearly, you want to get back to Boundary Creek."

"I grew up here. I graduated high school in Challis. So for me, the whole thing kind of comes full circle. I moved to Pocatello and didn't really like it. Too crowded. Now I'm going back to the Forest Service and moving back to Challis. That's better than 'okay' with me. That's great."

"Your folks worked in the mines. Why didn't you go to the mines?"

"I think because I watched, and saw it flip-flop a lot. My parents would go from being fully employed to looking for other work. And I wanted something more permanent. And my work on the river matters… it's not like I'm just out there to catch some rays or drink some gin-and-tonics. This is something that is absolutely a part of me."

"The whole spin of the world is in this direction," I predicted. "People need to be more disconnected."

"When Dad and I would take trips," Ted remembered, "we took a lot of doctors, and they told us it was medicine for doctors. A lot of people don't know about these places, where they can get disconnected. And maybe they're scared of it. They want the wilderness, but want their phones, their comforts. Once they get away from the cars and hotels, they slow down. And there's a benefit there."

"Andy was selling that a long time ago," I observed. "I mean, think about it. That was almost 80 years ago, and it had the same magic."

Ted sat up in his chair, and looked at me as if I had accidentally uttered one of the great secrets of his career.

"It has its own magic, Pat. Jack Bills, who was the Supervisor here for my first ten years, he used that river as a meeting place. Almost like a mediator. So he got the fish people, the environmentalists, and the miners – we took 'em down, sit around the camp fire at night, and discuss their problems. One thing about it – they get pissed off, they could walk down to the end of the beach, but they had to come back. They couldn't get up from the table and walk away. And that's how he got that thing started. That's how he got them working together. I took more than one group down that he was trying to massage. He got criticized for it, of course. But I stood up for him… I stood up to those people. 'Oh, he's just out there playing', they'd say. And he was enjoying the river, sure; he liked the river. But he was using it as a tool. There's all kinds of peace out on that river, and Jack knew that."

There was a difference between knowledge and wisdom. Even when people can't articulate the distinction, they recognize it. Lots of people on the river might say Ted's value was in his knowledge of the river and the wilderness, but it was the way he spoke about it that established his leadership. He was the best man to manage the Middle Fork because he was able to speak on its behalf with credibility and conviction.

Shared Joy

Cool nights were appreciated as June melted into the hot days of July. Heat shimmered off the highways in the high desert. Most of the people on the road were visitors from out-of-town, pickups that pulled travel trailers or toy trailers loaded with four-wheelers and dirt bikes. Motorhomes trundled up the two-lane blacktop, and cars piled up behind them and waited patiently for passing lanes. The water level had dropped in the last few weeks and it was prime time for Middle Fork adventure. It was perfect weather for splashing down the river in the heat of the day and relaxing by a campfire in the cool air of a mountain evening.

I pulled off the highway to watch some little birds mob a raptor... an eagle, as it turned out. It was always a curiosity to me, that a bird with an eagle's tools and temperament would tolerate that harassment. Mobbing was most common in spring, when baby birds were prey to a wide variety of predators on the ground and in the air, and their parents were focused on protecting them. I felt that was the reason the eagle put up with the smaller birds' crowding, literally flying within striking distance; the eagle knew they were protecting their young. Maybe he was between feedings and didn't want to waste the energy, or maybe he just wasn't bothered, but the eagle didn't fight the pests. He flew with the little birds like a shark swam with remoras. If you didn't know better, you'd think they were friends.

There were so many eagles along the river that locals sometimes took them for granted. As if it was no big deal to see them sitting in a tree, up high on a dead limb with an unobstructed view of the water. Sometimes in the summer, a tourist might pull over and walk back to the river, and the eagle might accommodate him with a photograph. Visitors could drive along the Salmon River for hours without realizing how close they were to wilderness. The eagles knew they were only a few minutes away.

I made sure to visit the wilderness a few times while I was in Challis. I spent a few weeks roaming around the backcountry that enchanted Andy Anderson, and camped in the Bighorn Crags for most of July, right after the winter snows melted out. Based on the articles I'd read that Tammy shared with me, I felt it was still the paradise it was seventy years before. Skirting around Heart Lake, I hiked over the pass to Terrace Lakes, down Waterfall Creek to the Middle Fork. Once I reached the river, it took a couple of days to hike up to the ranch on Loon Creek where Andy Anderson met Frank Swain, and took his historic first ride past Tappan. Loon Creek was halfway down the river; I knew it as a home. I lived and worked there for two summers irrigating and wrangling in 2014 and '15. Ted worked there as a teenager in the 1930s. I made camp on the hill that overlooked the ranch at the confluence of Loon and the Middle Fork.

Ted told me stories about his days on the ranch. He told me where they used to hunt out of the camp 'back in the day'. We shared the grist of backcountry history; all the stories everybody knew if they lived around there and listened long enough. Stories about places and their histories, the ever-treasured stories of the people that inhabited a place like the Idaho Primitive Area. He talked about working on the runway with Shorty, and irrigating the pastures the old-fashioned way. He had lots of good memories about Shorty.

Apparently, Shorty Waits was a memorable character. Ted was sixteen years old when he met the old hand, and he shared several stories about Shorty with me. They worked together putting up hay at Simplot and shaping the backcountry airstrip with a Fresno plow. It gave them plenty of time to talk.

Shorty claimed to have known Earl Parrot, the famous hermit who lived at the lower end of the river who was discovered during one of the trips made by Doc Frazier and the club in the '30s. Shorty told Ted he met Earl at a boarding house, a place they occupied at the same time. The owner of the boarding house liked to tease the young men that stayed there, and one night over dinner teased young Earl about not having a gal. Earl was embarrassed, got up from the table, and left the boarding house for good. It was the beginning of his famous hermitage, and Shorty told the story convincingly.

Ted's favorite story about Shorty took place at the Simplot Ranch. Shorty said his wooden matches kept disappearing, which was more than a nuisance as supplies were hard to come by and matches were needed to make fire and keep warm at the isolated backcountry ranch. Investigating their mysterious disappearance, he discovered a pack rat had been stealing the matches and stashing them near his nest in the rafters. Shorty attempted to shoot the rat and, in the process, accidentally started a fire in the cabin where he lived. On the roof to fight the fire, he jumped off when it roared out-of-control and hung a shoelace on a shake roof shingle. Dangling upside down by the shoelace, he pondered a solution to his urgent problem.

"I says to myself, says I, 'I got to cut that string!'"

"That's just how he talked," Ted chuckled softly when he quoted Shorty's words to me. His face smoothed out as he remembered his early days on the Middle Fork. I asked him once about any dark days on the river, and he said after considering the question that he could not recall dark days. Tough times and lots of hard work, maybe, but nothing that weighed heavy on his mind.

He leaned on the armrest as he got comfortable in his chair, then straightened up and said, "I'll tell you one of my favorite stories from the olden days. Dad would have a base camp at Hospital Bar. He could ford just downriver from the hot spring, and he hunted up on the other side. He and Bill Sullivan both like to hunt on the other side of the river. Bill worked out of Stanley; him and Dad were friends, both big storytellers. People would stand around and listen to them trying to out-do each other. Well, Bill was hunting out at Sheep Creek that year. Do you know where that's at?"

"Yes, sir."

"And he had his camp up Sheep Creek, where the trail goes up to the ridge and over into Brush Creek. What a lot of people don't know is how close things are together if you go the other way. Not far to Hospital Bar, Norton, and all that. So they got up in there one morning and it was foggy as hell, and I don't know if they had the guests with them right then. But anyway, they started bugling. Started callin' in the elk. This one would call, and that one would answer. They didn't know they were there at the same time, and they bugled each other in! Oh, they had a laugh when they found out. One of them had a bottle, so they sat down and had a drink. They bugled each other in through the fog."

Ted laughed, and I laughed; we both knew how fun it could be in the backcountry. Where you forget for a while the part you play outside, and relax in the communal comfort of just being human. Everyone that divests themselves of the digital world and invests in Time Alone shares that feeling of kinship with one of the best parts of this life, the joy of being free in the wilderness.

That joy was not anyone's private property; in fact, it was best enjoyed when shared with others. As Andy did, which was why he started the Bar-X way back in the 1930s. It was the reason Ted returned to the river after trying other things, and stayed in service as River Manager for the rest of his working life. The Andersons made sharing the Middle Fork and the Frank Church Wilderness a family vocation; Andy sold the idea, and Ted made it work. They reflected and exemplified the values of their community, the beliefs and standards of conduct that made this country great. Theirs was a lasting contribution to the river and the wilderness, and to those of us that benefited, either directly or indirectly, from the Anderson legacy.

Manufactured by Amazon.ca
Bolton, ON